Multimedia Communication

Multimedia Communication

Andy Sloane

Senior Lecturer
CoNTACT Research Group
School of Computing and Information Technology
University of Wolverhampton

THE McGRAW-HILL COMPANIES

London · New York · San Francisco · Auckland
Bogotá · Caracas · Lisbon · Madrid · Mexico · Milan
Montreal · New Delhi · Panama · Paris · San Juan
São Paulo · Singapore · Sydney · Tokyo · Toronto

Published by
McGRAW-HILL Publishing Company
Shoppenhangers Road, Maidenhead, Berkshire, SL6 2QL, England
Telephone 01628 23432
Fax 01628 770224

British Library Cataloguing in Publication Data

The CIP data of this title is available from the British Library, UK.

Library of Congress Cataloging-in-Publication Data

The CIP data of this title is available from the Library of Congress,
Washington DC, USA.

McGraw-Hill

A Division of The *McGraw·Hill* Companies

1 2 3 4 5 CUP 9 8 7 6

Typeset by the Author.
Printed and bound in Great Britain at the University Press, Cambridge
Printed on permanent paper in compliance with ISO Standard 9706

In memory of Dr Hugh Court

However entrancing it is to wander unchecked through a garden of bright images, are we not enticing your mind from another subject of almost equal importance?

Kai Lung's Golden Hours. Story of Hien
Ernest Bramah 1868-1942

CONTENTS

PREFACE

The multimedia "revolution" has received a lot of attention in recent years but has not really given much more than a brief insight into what is possible, even though there are now a considerable number of products, both hardware and software, that use the word multimedia to describe themselves. Alongside this is the growth of computer networks which are now becoming available as the "Information Superhighway". Together these two developments open up a much wider field of possibility for business, commerce, industry, and also education, healthcare and leisure. These developments link together to form a branch of knowledge which is known as multimedia communication. What this means for the future of business, education and individuals will be explored in this book.

The book is about the systems and techniques that can be used to develop better communication using multimedia. It investigates what is at the heart of this convergence of both the multimedia approach and the increasing availability of network connections. It also focuses on a number of typical applications that benefit from the application of multimedia communication technology and systems. These are taken from the application areas that are in the forefront of multimedia system development and also from areas that are not yet considered as priorities.

The intention in writing this book has been to create a stand-alone text that can be used in a number of different ways. Firstly as a standard textbook to accompany a series of lectures, secondly as a self study text in which further material is provided by on-line access and thirdly as a book for the general reader that covers the wide area of knowledge that is encompassed by this subject. It is, therefore the intention that the text should be, as far as possible, complete with reference to outside sources where topics are not covered in depth. One such source is the author's previous text, *Computer Communications* also published by McGraw-Hill that forms a useful introduction to the communication aspects covered in this book. The provision of on-line resources to illustrate the points raised in the book is also a consequence of the networks that the book describes!

Because of this range of coverage much of the material at the beginning of the book is of a fundamental nature and this is intended as a background primer for those coming to either of the main themes for the first time. Consequently some of this will be familiar to readers who have studied these subjects before. It is hoped that the treatment will, however, be new and even well-read students will gain some new insights! In this section on fundamentals a main theme will be developed; that of the nature and role of multimedia information and its communication.

Please enjoy the book.

Andy Sloane
23 September, 1996

SUPPLEMENTARY ONLINE INFORMATION

Various topics throughout the book cannot be sufficiently explained by text and pictures alone and so an online site has been developed to contain this extra material and some up-to-date references. Because of its nature this is being kept on a World Wide Web site which can be accessed by browser starting at the publisher's URL

http://www.mcgraw-hill.co.uk/

The links will be found with the entry for this book in the online catalogue.

While the author and publishers will endeavour to keep this information available and current, no guarantee is implied for the continued provision of this supplementary material.

1

COMMUNICATION

Summary: Overview of multimedia and communication, possible uses, current developments: the relationship between human communication and multimedia communication. Models of communication: human, data and information communication.

1.1 INTRODUCTION

The ability to communicate has long been thought to be the fundamental basis for human dominance of the world. It is, however, not as clear cut as it may appear at first sight. There are many different ways to communicate between two individuals, whether they are humans or some other animals. What has developed through evolution are systems of communication between members of the same species that are suitable to that species and often allow only the intended recipients to understand the information that is being transmitted. So, for example, although we may consider human communication to be the most complex example of this type of information transfer, we cannot really claim to understand the nuances of the communication system used by whales in the oceans. What is clear, however, is that all forms of communication have the same goal, even though it may be that the scale and complexity varies between different systems. That goal is the transfer of information between the source of the communication and the receiver. Therefore, a conversation between two people, a television broadcast, and a scent marker left by a dog are all examples of different types of communication. Each of these examples has different amounts of information being transmitted, but each method is usually appropriate to the transfer of relevant information and the system being used for its transfer.

To expand on this example further may clarify what is important. A television broadcast is a device for sending information between a central broadcasting organisation and a large number of people. Mostly, the information that is transmitted takes advantage of the medium that is used, even if it doesn't do so for all the time the channel is in use. The television medium is a combination of sound and moving images which allows two separate channels (sound and vision) to be used to transfer the information in a broadcast fashion to a large number of passive recipients along a one-way television channel. The nature of the information transmitted can be very diverse, because it relies upon the spoken language and visual images to encode the ideas of the broadcasters for transmission over the system. If this is compared to the scent markers left by dogs and other animals which have a limited information content (basically identifying the position of the marker as territory belonging to the transmitter of the marker scent), it is clear that there is a similarity in the process but the content of the communication, i.e. the information contained in the communication is at different ends of a very wide scale.

What is important, then, is to distinguish between the information content and means of transfer. The message and the medium. The message is the information that is sent and the medium is the means of transfer. Humans have developed a number of media for communicating information in different settings and scenarios. Before the advent of the printing press, the only means of transferring information to large groups of people was to use public meetings and word of mouth. Then the use of written communication allowed newspapers to carry information to the masses and eventually radio and television have supplemented the written word with spoken and visual communication channels for broadcast. Finally, the latest technological development is to incorporate all these methods of communication together into a multimedia form of communication that allows different types of communication to be used for different types of information and enables an appropriate means of transfer to be used.

This new development of multimedia communication has been made possible by the increasing power of computer systems and the continuing pervasiveness and diversity of the communications networks that are now part of everyday life for many people throughout the world. However, what is not always clear is how this development will be used and how it will enable the transfer of information between its source and the intended recipients. This book will look at the two areas mentioned in the last paragraph (medium and message) as different aspects of the communication process. Firstly the media that are used for electronic communication will be described and then the use of multimedia will be covered and each of the component parts that are used will be discussed. The rest of this chapter will look into various aspects of the communication process to give a framework for the later material.

1.2 INTRODUCTION TO COMMUNICATION

Communication is an intricate process that has many factors that influence its usefulness. These factors will vary according to the medium used and the type of information being transferred. For instance, it is possible to communicate using speech between two people in a quiet room without any misunderstandings taking place. If the same conversation were to take place on a street corner there would need to be some alteration of the volume to take account of street noise and often the recipient would need to ask for repeated words and phrases to completely understand the information being transferred. Noise, however, is only one aspect of the communication process, as will be seen later.

What is often taken for granted is that the means of communication that is used is the appropriate means of transferring the information that is required. As an example, consider a tennis match as a source of information. If the intention is to communicate the score then either a written or spoken communication will achieve this fairly efficiently, but if the intention is to give some idea of how each player played each ball then a spoken or written version becomes a little unwieldy. This is where a video version of the tennis game will prove most useful. Having said that, if the only interest for the recipient of the information is the score then the video version is a waste of the medium and a more appropriate form of information should be used. This will be expanded later in Chapter 3. It is worth noting, however, that there are often reasons for using a particular medium that are not only related to its technical appropriateness to the task. For instance, an interview between a television presenter and the leader of a political party before an election would, most likely, command a prime time slot on a television channel, even though the two participants would only communicate using speech and there would be no real information lost by having the interview carried out on a voice-only radio channel. The determining factor here would not be the technical needs of the communication but the likely audience for the communication.

It is clear that the process of communication is complex and there are a number of factors that influence the choice of channel that is used and the message that is transmitted. Some of these are shown in Table 1.1.

Table 1.1 Factors influencing choice of communication channel

---

Availability of channel
Availability of equipment
Ability to use channel
Ability to understand message
Economic considerations
Time problems

As examples of how these factors can influence the communication process, consider the following

1. How would you contact a friend who is working in the jungles of South America where there are no radios, or telephones available?

There is no available channel or equipment.

2. How would you communicate with someone who is deaf, and only communicates with you in sign language, when you cannot see them?

You would need to be able to see them and be seen by them to be able to transfer messages. So the ability to use the available channels is missing.

3. How do you reserve a room in a hotel overseas and the person answering the telephone speaks only Italian and you speak only English?

You have a channel but no means of using it and the recipient cannot understand the message.

4. You want to be able to make and receive telephone calls while travelling about the country.

A mobile telephone would be the only means of communication to allow this flexibility but if it is too expensive to use it may not be an economic proposition.

5. You want to make a business call across the world. You are in France, the recipient is in Australia.

The time difference in countries this far apart can be a problem. There is a considerable difference in actual time.

These examples, although relatively simple, do illustrate the problems that can occur in trying to use communication channels. There are a number of these factors that govern the use of channels and if any of them are relevant to the use of a particular communication channel, the transfer of information will be impaired or impossible. What will be seen later (in section 1.5), is that there are a number of fundamental rules which govern communication and each of the factors listed will be included within these rules. There are also internal rules that govern how communication takes place and these will also be included in the later discussion.

Although, this book is about multimedia communication, which relies heavily on the underlying infrastructure of computer networks and the telecommunications systems around the world, it is worth remembering that the means of communication does not alter the basic premiss that the act of communication is as a means of transferring information between people (mainly) and this is the basis behind this book and, indeed, behind multimedia communication. So, although the technology is new and the ability to use it effectively is not yet perfected, the principles behind the use of these media are no different from those that have governed communication between people over the centuries.

1.2.1 Human communication needs

As has been mentioned above, humans communicate in a variety of different ways. The main methods being through speech and writing, although various sign languages have been developed to suit different circumstances. There are, in fact, many ways in which limited information can be communicated between people and the use of a coding that is understood by each of the communicators is a fundamental part of the process. For instance, in using speech to communicate we are encoding the ideas that we have, that is, the information we wish to transfer, into a common encoding that can be decoded by the recipient. The encoding is learnt early in life by listening to adults speak and the encoding that is used is some form of language. This is often a standard language such as English, Italian or German, but can also be a minor variation of these, a dialect. Communication can be easily carried out using a standard language but difficulties can occur when a dialect is used.

The other main form of human communication, the written language, is now much more of a standard than it was in earlier times and more so than the spoken word. A number of good examples can be found in old manuscripts, where words often use different spellings from the current standard although the meaning is the same. The spoken version of a standard written word, in different dialects, may also vary from the standard pronunciation.

There is, therefore, a large degree of standardisation in the human aspects of communication. This is inevitable and useful. Firstly, it allows large numbers of people to communicate directly, without interpreters, and it saves a lot of confusion. This standardisation is a direct consequence of communication itself, since the need and ability to communicate will lead to a greater standardisation in time. When countries were small and people relatively isolated then the local language was all that it was necessary to know. However, now that communication takes place on a global scale there are, increasingly, moves to standardisation world-wide. One example of this process is the use of English as a common language at many international scientific conferences and on the Internet.

However, this does not change the basic human communication process. The spoken and written word are still the main forms of communication between people even though there is now more variety in the methods of using these two means of communication. The use of computers to communicate has added a new dimension to the accepted means of communication. What will be clear is that the traditional means of communication will continue to be used although they will be greatly enhanced by the use of computers and other devices in the coming years. This will aid communication and allow more users to communicate more effectively than has been possible with the more traditional means of expression available to them.

1.2.2 Machine communication

Computers have made many changes to the way many things are accomplished. In the last 50 years the methods of doing simple everyday tasks have changed considerably. Consider the now common process of withdrawing cash from a bank ATM (Automatic Teller Machine or cash machine). These have only been in general use since the 1970s, before that time the usual method of withdrawing cash was to use a cheque over the counter. The transaction that used to take place between two people is now an interaction between a person and a computer system. Even where there is an apparent person-to-person interaction there is often machine-based information that is used to order the communication. For example, if a customer makes a telephone enquiry about an order to a company it is probable that the information used for the communication will be held on a computer system, even though the communicator of that information is a person.

Going one step further, there are now direct machine-to-machine links between computers in different organisations and some information is automatically transferred between these systems to conduct regular business. This use of machine-to-machine communication is most widespread in the EDI (Electronic Data Interchange) networks that are used for the conduct of business in many sectors of industry and commerce, most notably in the automotive industry where manufacturing is now largely dependent on the computerised ordering of parts and scheduling of deliveries.

It is possible to think of computers as a different kind of communicator. After all, the language that is used to communicate between two computers is very different from spoken or written language that is familiar to humans. Computers use binary digits (bits) to send signals and messages. If these messages are sent between two computers, then they do not have to comply with human forms of communication. They do, however, need to abide by the general rules of any communication which will be discussed in section 1.5. In fact it would be very wasteful of the communication channel to include information that is only of use to human users when the communication is solely between machines. This is, however, unusual as most computer-to-computer communications will eventually be used by a human operator, or user somewhere in the chain.

What was clear in the last section is that human communication needs are for the basic system of language to be able to be used via speech or writing so that people can communicate. This is a far more complex problem than being able to transmit two different bits between computers. Language consists of a complex set of sounds or written symbols each of which are used in a multitude of combinations to produce sounds or words with some information content.

So what is the connection between computer communication and human communication?

Put simply, the rules of communication are the same but the difference lies in the different symbols used to communicate. Computers use bits, people use words. However, it is possible to use computer communication to transfer information intended for people. What has to be done, is that the information that is encoded in speech or writing needs to be encoded again in bits. i.e. it needs to be converted into a digital signal that can be directly used by the computer systems. This usually makes the information less compact than that used between computers but much easier to understand for humans!

Different types of communication have different characteristics. Each medium that is used, requires a different method of encoding the information that is transmitted. Speech uses sound, writing uses alphabetic symbols. Each of the systems used by people to communicate can now be encoded to be transmitted by computer. Each of these types of communicated information needs to be encoded as bits, and there are a number of ways of achieving this. These different types of media used for communication are discussed in the next section.

1.2.3 Media types and use

The process of communication has given rise to many and varied forms of transmitting information. Principally, as mentioned in the previous sections, speech and writing have been the preferred means of communication for many centuries. There are, however, many forms of communication that use image or moving pictures (video) to transfer a message. How these are used has always been a point of discussion but there are now systems that can allow many different types of communication to be employed. The main information types are shown in Figure 1.1, in the order of size for a typical piece of information.

As the complexity of each communication type increases so does the size of typical objects of the type being considered. Text is the simplest medium to store and transmit and can contain a large amount of information in a relatively small space (consider how much information is in this book and the space it occupies). The other media increase is in terms of the complexity of the machinery needed to exploit them and in the techniques used to store and transmit the information.

Video, is the most complex and, potentially, the most massive of all the media. This ranking of the different media types will also be expanded in more detail when the digital forms of these media are considered in Chapter 3. There will, of course, be some instances of each media type that fall outside these boundaries, but in general the progression is straightforward.

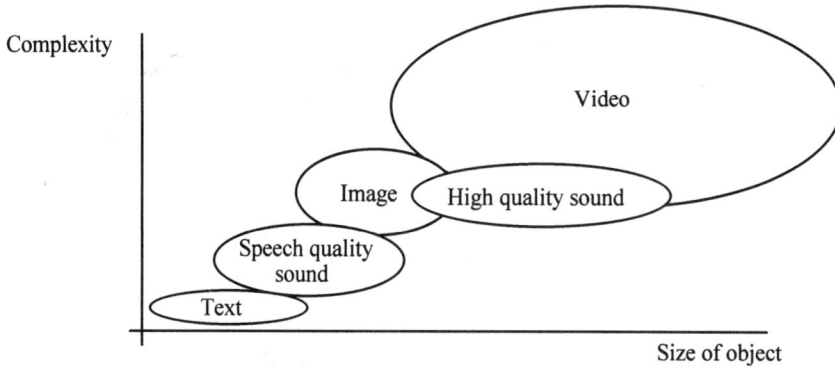

Figure 1.1 Media types and complexity

The use of each media type is also relatively standard. Text can be used for many communications and is still a major force in communication, as it can encapsulate complex ideas and be used as a storage and transmission medium relatively easily. Speech quality audio is used where it is only necessary to relay the spoken word, other more complex sounds (such as music) can use a higher quality of sound processing. It is, however, worth noting that the human ear has physical limitations that will not be overcome by using even higher quality sound, so there are limits to the usefulness of increasing the quality level.

Often to provide a simple means of communication of ideas a diagram, or illustration, is used. In this case an image can provide a useful way of transmitting this graphical information. Finally, if the need is to see an event, then video would seem most appropriate.

What has happened over the many years in which storage and transmission systems of these various media types have evolved is that there has been an increasing use of the most complex technology to aid communication and an increase in the use of all forms of communication using any medium. What the digital communication age has allowed is the convergence of all these forms of communication, the inclusion of one with another and the interaction of users with a variety of media through the use of digital communication and the computer. How the communication infrastructure has developed will be discussed in Chapter 2. The next section will briefly introduce the ideas behind multimedia communication and the use of different media types together.

1.2.4 Multimedia communication

The combination of different media types into a single coherent object is what is now commonly called multimedia, although there are often confusing uses of the terminology in everyday use that often mean much less. The use of video sequences with audio soundtracks, text and graphics overlay and still images is what is generally defined as the multimedia paradigm. This allows all communication types to be used either simultaneously or within a particular sequence to allow best use of the technological ability of whatever devices are being used.

There are, however, strong defining links between the computer, digital storage and communication and the multimedia concept. The ease in which computers can manipulate digitally encoded objects allows the information technology devices to become *multimedia* information technology devices with a little additional hardware and software. There are, of course, certain limiting factors on speed and processing power which determine how useful a particular computer will be at handling multimedia information. For instance, it is no use using video recorded at 30 frames per second if the playback device can only handle 10 frames per second!

Where this multimedia capability is now being used effectively is in the communication of ideas between people, that is in multimedia communication. The ability to use one channel divided between a number of different media types is now a possibility. Therefore, the communication between the transmitter and the recipient of the communication can now use the most effective media for each different communication need during a conversation or communication session. The liberating effect of this opening of the communication channels has had a profound effect on computer communication during the 1990s. The ability to switch between media types has allowed the communication process to be aimed more directly at the use that is envisaged by the actual users. For instance, if an image is an integral part of a document its inclusion will not now be a problem since there are encoding techniques that can deal with this mixture of media types, and best use can be made of the communication process.

It is, of course not always as easy as appears to be implied by the last paragraph. There are many problems that have a negative impact on the real communication environment, amongst which is the choice of the various standards that are available. The need to have an agreed procedure for each end of the communication process is one of the underlying rules of communication that will be defined later. This is doubly true of multimedia communication where any slight alteration of the standard formats can mean the difference between a good communication and none at all.

One of the first uses of multimedia in communication has been the development of personal communication tools. These most often include video and audio communication as a background to other used media, either text and graphics in a whiteboard-type system or just text overlay. Figure 1.2 shows a screen from one example of this type of tool.

Figure 1.2 Video conferencing screen with text

As can be seen from Figure 1.2 the use of text is as an additional communication medium to the audio and sound although in this particular application it preceded the sound capability and was the only means of communication apart from the screen picture until the sound capability was developed. There are now a number of commercial tools available on many different platforms to achieve this type of communication, and more sophisticated types of multimedia communication.

To analyse the situation further needs some basic definitions and analysis. Firstly the nature of information itself needs to have a structured approach to multimedia and secondly the communication process needs to be defined and analysed in detail so that the needs of the process are not overlooked. Section 1.3 looks into information, section 1.4 is dedicated to the process of communication.

1.3 INTRODUCTION TO INFORMATION

Although a more detailed discussion is contained in Chapter 3, it is important that the term *information* is given some basic meaning at this stage. Firstly, information and data must be differentiated. There is often a misconception that data and information are interchangeable terms. In this book there will be a distinction drawn as follows.

Data is the raw quantity, whereas **information** is the same data plus some meaning.

This meaning will be drawn from the context of the data. This could be and often is defined outside the communication or storage of the data. For example, Table 1.2 demonstrates the differences between raw data and the information contained within it.

Table 1.2 Data and information

Data	Information
2343256	Telephone number of a friend
29WS157QQ	UK House number and postcode
2894251992	Product code of a compact disc

It is clear that without the contextual clues the data may not be easy to decode and its original meaning may not be apparent. There is, however, a good reason for differentiating in this way especially in a communications context. As will be seen later, the communication process defines a context in which the data can be transmitted and the original information recovered from the process. Likewise there are sometimes flaws in the communication process that result in information being lost because of either data loss or faulty contextual encoding, resulting in an incomplete recovery of the original information.

This above analysis shows that information can have a particular meaning depending on the context, what is also possible is that a large amount of data can have many derivable meanings contained within its data.

1.3.1 The structure of information

Data can be viewed as a series of symbols. Depending on what the encoding system is that is used for the representation there may be as few as two different symbols as in binary digits or many more. Table 1.3 shows some of the common symbol sets that are used. There are many more in everyday use in different countries of the world and in different information environments.

Table 1.3 Data symbols

Name	Symbols
Binary numbers	0 1
Denary numbers	0 1 2 3 4 5 6 7 8 9
English alphabet	a b c d e f g h i j k l m n o p q r s t u v w x y z

The number of symbols that are used can vary, even between similar encoding systems. Table 1.4 shows the differences between the English alphabet and the Greek alphabet. These are not great, especially as they are both codings that are used to represent spoken languages which have similar sounds. There are equivalent symbols but the Greek set is only 24 symbols compared with the 26 in English.

Table 1.4   Similar symbol sets

Name	Symbols
English	a b c d e f g h i j k l m n o p q r s t u v w x y z
Greek	α β γ δ ε ζ η θ ι κ λ μ ν ξ ο π ρ σ τ υ φ χ ψ ϖ

This shows the variety of different symbol sets that have been used over the many centuries that people have been communicating. All of these are inventions that are designed to be used to represent the quantities and values in any communication. Any concept can be encoded in a number of ways for transmission to different recipients. The list below is a set of different encodings of the same concept

Tuesday
Mardi
Martedi
Dienstag
Martes

These different encodings of the same concept are all equal, and they are all used in the particular context of the language to which they belong. So, given a different context it is necessary to encode a concept differently to be able to communicate it even though the symbol set used is (more or less) the same.

These different factors, i.e. context and symbols used have an obscuring effect on the information contained within a piece of data. The context needs to be known so that a full decoding can be achieved. This is not usually a problem, but when machines are used the definition needs to be fairly exact so that no errors occur. What the context of communication provides is a structure to the information to enable decoding.

When a binary encoding is used, as in computer systems, the structure of the information, and therefore the context of the communication, is provided in a number of ways, sometimes simply by position and sometimes by using conventional language techniques. For example, a simple message between two computers might contain the following data:

234556671+25+23456673+100+23456688+10

This could actually mean almost anything, but if the context of it being part of an order for parts, and the longer numbers were known to be part numbers and the smaller numbers quantities, then the message is quite clear. On paper the message could read:

Order

Part Number	Quantity
234556671	25
23456673	100
23456688	10

There is, however, much more information contained within data than these simple examples can show. As an example, a digital video sequence is a long bitstream, consisting of many bits for each second of the video sequence. Contained within this will be a series of pictures and an audio soundtrack which together contain a record of some event. The information contained in the bitstream enables the event to be reproduced as a video sequence. The bitstream may also have some derivable information that is not explicitly contained within the bit stream.

Consider the example mentioned in section 1.2 of the tennis match. A digital video could be produced of the match containing a complete record of the play, this would consist of many millions of 1s and 0s, but from this could be derived the score of each set and game and the various pieces of statistical information that are of interest (e.g. the number of aces served by each player, the number of double faults, etc.). So, contained within a large number of bits, is the information required to produce the summary information. Although this is, perhaps, an extreme example using video as the source, there is in general a lot of information that can be extracted from data. What is also clear is that where more data is available, more information can extracted.

1.3.2 The nature of information in communication

The need to have effective communication places some restrictions on the actual process of communication itself. It may be implied from the previous section that more data equates to more information and, whereas, this is generally true it may not be an efficient use of a communications channel to add more and more data to improve the quality of information received. This problem has been around since communication began, i.e. to have an effective communication using an efficient means. The added problem that occurs when information is being communicated is the possibility of errors occurring in the process.

So, there are two opposing forces determining the communication process. One is the need for accurate information to be received, which may require extra data to be sent. That is, it may need some essentially redundant data to ensure accurate reception. The other factor is the need to reduce, as far as possible, the data sent to minimise the use of the channel. This may be because of cost or some other factor. It may also be counter-productive to include too much extra information in a communication as it can, in certain circumstances, obscure the relevant information.

One other factor that may be worth noting is the need to have information transmitted within a particular time. If the information is intended to be used in real time, as may be the case with digital sound or video, it would be inappropriate to include much extra data if it removed the capability of the channel to transfer the data in sufficient time for it to be replayed as intended. This is often a limiting factor in multimedia communication where sound and video are components of the communication information being transferred.

1.4 MODELS OF COMMUNICATION

The previous few sections have detailed some of the problems involved in communication. There has already been a brief discussion of encoding as a means of utilising channels for communication. This will now be extended into a more general model of communication where the transfer of information is accomplished by the various coding techniques that are needed and used in everyday communication and information transfer.

1.4.1 General model of communication

As has been stated already the purpose of all communication is the transfer of information from one information user to another, as in Figure 1.3. This shows the perfect situation where a piece of information is transferred from User A to User B and is unchanged in the process.

Figure 1.3 Information transfer

This is not always the case, as has already been shown in various examples. The transfer can have errors and the interpretation of the transmitted data can result in different information at the two ends of the communication process. The rules that govern communication will, to a certain extent circumvent this problem, but there are still areas that need close scrutiny to ensure a good transfer takes place. A frequent problem is that shown in Figure 1.4. Here the information has been sent but has been interpreted differently, but not significantly so. (The difference is shown by the change of font and style.) This is a common problem and one that may not be easily solved.

User A   **User B**

Figure 1.4 Slightly faulty transfer of information

Another common problem is that shown in Figure 1.5. This is when the transfer process itself is disrupted and the information is lost or severely disrupted. Noise, as shown, is a common cause of this and efforts to eliminate or accommodate noise are part of any communication model.

User A   **User B**

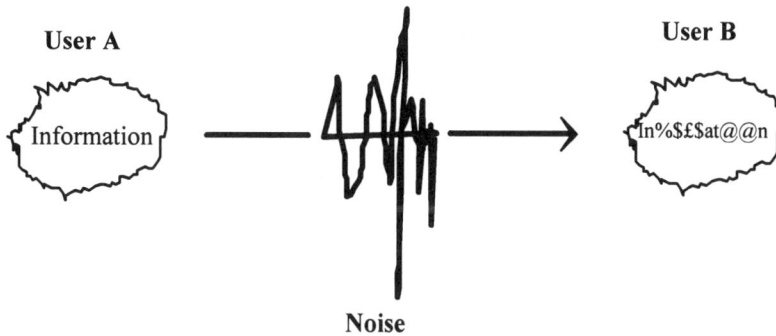

Noise

Figure 1.5 Noise affected communication

To allow for a solution to these problems and to enable users to use the channels available the information has to be encoded. The method of encoding will be determined by the available medium and the ability of the communicators to use it. If all these factors are in place then a simple model of the process is as shown in Figure 1.6. The information is encoded in some way appropriate to the channel, transmitted over the channel, decoded by the receiver and the process is complete.

This is, of course, a simplification. Mostly, communication needs more than one level of encoding and this multi-level approach has a number of benefits. Firstly, the final encoding can be directed at the channel. i.e. the final level of encoding can best fit the available channel. At the highest level the encoding can be made appropriate to the receiver of the information, and finally the intermediate levels can be made appropriate to the likely problems of the transmission. A more complete diagrammatic description is in Figure 1.7. This multi-level approach is further explained by means of an example in the next section. At this stage although the model is quite complex and can effectively allow many different types of communication to be described it does not deal effectively with multimedia communication.

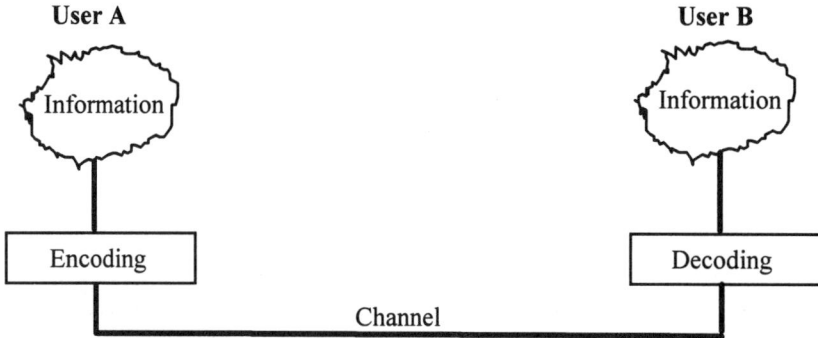

User A   **User B**

Information   Information

Encoding   Decoding

                          Channel

Figure 1.6 Process model of communication

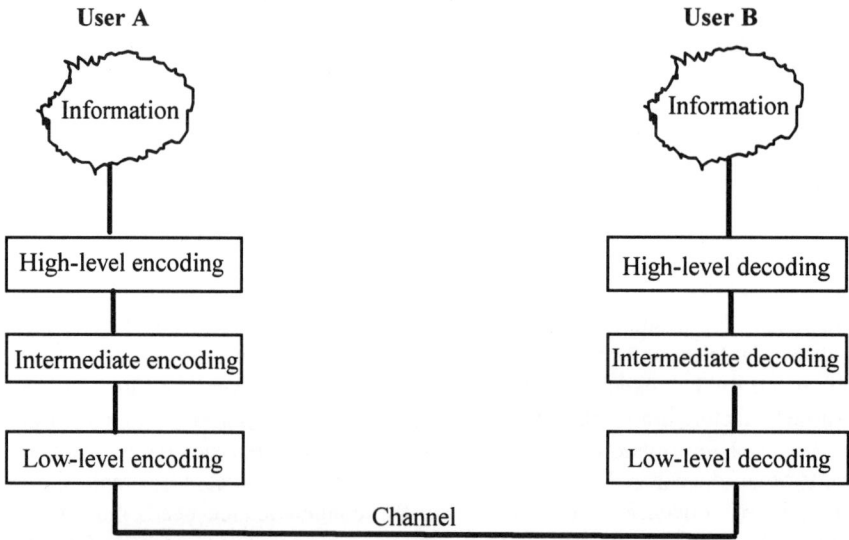

User A   **User B**

Information   Information

High-level encoding   High-level decoding

Intermediate encoding   Intermediate decoding

Low-level encoding   Low-level decoding

                          Channel

Figure 1.7 Multi-level encoding

This is mainly due to the emphasis, within the model, on single channels and single media types. Multimedia communication will inevitably involve more media types and although one channel is used the encoding will require an additional layer to incorporate different media within one channel simultaneously. This simple addition to the model for multimedia is shown in Figure 1.8. The different media used can all be incorporated into a single signal if they all digitally encoded. The various media encodings are mixed or multiplexed together to form the single multimedia channel.

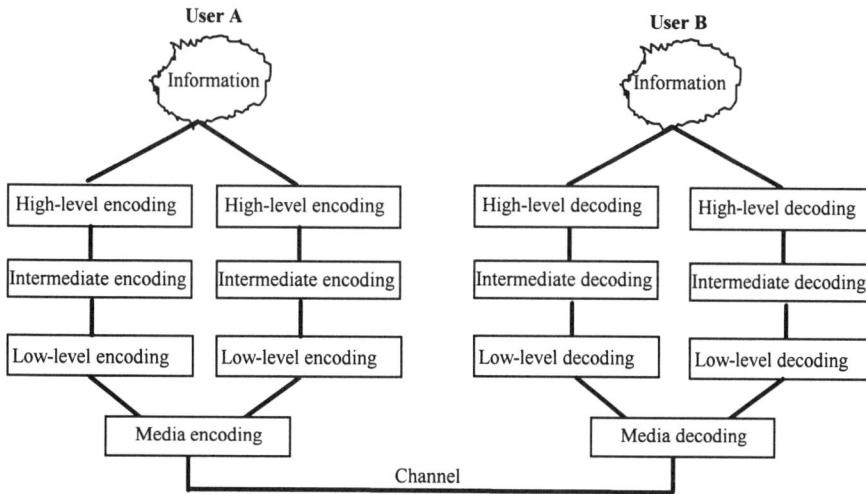

Figure 1.8 Model for multimedia communication

1.4.2 Examples from human and computer communication

Example 1

Human speech

The speech process is one that is common to most humans. The transfer of information is simple in theory, but may be problematic in practice.

The process involves the transmission of ideas (information) using the following encodings.

High-level The language of the communicators

Low-level   Sound

The channel being the air space between the transmitting person and the receiving person.

Example 2

An electronic mail message

One of the first and most pervasive of computer communication applications, it uses a number of encodings to achieve transmission.

High-level   The language of the communicators
Written symbols of the above language
Computer representation of the language symbols (e.g. ASCII code)
Low-level   Electronic signals at two levels (for 0 and 1 bits)

Example 3

A multimedia document

As an example consider the combination of a digital video conference with voice sound channel and text overlay.

Video encoding

High-level   Information encoded as series of images (25-30 per second)
Series of images encoded digitally
Low-level   Digitised into bits

Audio encoding

High-level   Language of user
Sound
Low-level   Sampled and digitised into bits

Text encoding

High-level   Language of the user
Representation as symbols
Encoding of symbols into standard computer code

Low-level   Bit encoding

All three encoded signals then need to be multiplexed into one bit stream for transmission. This can be done in a number of ways.

1.5 PRINCIPLES OF COMMUNICATION

In the preceding sections the various aspects of communication have been outlined. Included in this discussion have been a number of pointers to the important features of communication that provide a set of principles that govern it. These principles are common to all types of communication including both human and machine-based communication. They provide the rules that govern the transfer of information so that it can be understood upon reception. Any misunderstanding of the intended meaning of a communication will be due to some failing in one or more of these aspects.

1.5.1 Generic principles of all communication

There are six principles to consider, each of which needs to be defined either explicitly or implicitly for a particular communication task. The principles are as follows:

1.   **Standards** The use of particular methods for any aspect of the communication process (e.g. the use of a particular language or specified coding technique).
2.   **Protocols** Agreement on the framework for communication which may also include the specification of standards for this aspect.
3.   **Error control, redundancy and accuracy** To control errors requires that there is some extra information in the communication that enables either the detection or correction of a problem. This redundant data allows recovery of the intended information from the communication. This provides an accurate transmission of the original information.
4.   **Channel** Communication takes place between end users via a channel. There can more than one channel in use for one communication session, or more than one mode of communication taking place on a single channel as in the case of multimedia communication.
5.   **Context** The context within which a communication takes place helps to define some aspects of other parts of the communication. For example, in a speech communication between two strangers in an English town it would be assumed that English would be the appropriate language to use.
6.   **Coding (Encoding and decoding)** Accurate and unambiguous coding is essential for accurate transmission of the information. This may require a number of separate levels of coding to achieve a reliable scheme that can use the available channels.

1.5.2 Standards

While all six of the principles listed in section 1.5.1 are important, the position of standards is unique in that they can be incorporated in all the other principles in one way or another. There are frequently occasions when the complete communication process can be defined by the standards that are being used in it. This allows a much simpler definition of the parameters of the communication than would be possible without a pre-defined standard method. This is best illustrated by an example.

Example ·

Some standards used in the transfer of an electronic mail message

- The message itself will contain text in a standard language.
- The format of the message will need to be aligned to one of the generally available standards.
- The transfer of the message over a network will require standard protocols and signals.

This shows some of the areas that have been standardised for ease of communication in a particular environment. In some more common settings the standardisation may be taken for granted (e.g. in everyday speech). In other, less familiar communication paradigms, such as multimedia communication the standards will need to be defined in detail to enable an effective communication between the interested parties. Some of these standards will be defined in Chapter 3 in relation to different media and multimedia communication.

1.6 USE OF DIFFERENT CHANNELS AND MEDIA IN COMMUNICATION

The digitisation of communication has had two effects. One is to require that all means of communication become digital, the second is the increasing need to be able to provide a means of transmission for digitally encoded information, in ever increasing quantities.

To give some idea of what is being achieved it is necessary to understand some of the basics behind computer communication. Chapter 2 is dedicated to this requirement. What is now important is that all communication media can be digitally encoded and this means a simplification in the transportation of information that is not possible without this standardisation.

The different types of media that are normally included within the multimedia scope are:

Video	
Audio	Speech
	Sound and music
Image	Graphics
	Picture
Text	Free format
	Structured information

Individually these are all familiar and have been in use for many years, and in some cases centuries, There are now, however, possibilities that were previously difficult or impossible. These have been made possible by the use of digital technology to encode, store and communicate information between users of computer systems. There is also the possibility of using these digitally encoded forms of information in ways that have previously been useful for various purposes. For example, a common method of displaying information to an audience is via a slide show, with a commentary. This is easily accomplished using multimedia as it involves the use of images and a synchronised audio soundtrack. Likewise, the picture album can be replaced by digitally encoded images with a text annotation. The various components of each of these digitally encoded forms can be combined together in both ways that have been commonly used before and new sequences and connections that are still under experiment.

1.7 CURRENT TRENDS AND IMPACT ON SOCIETY

There are two main threads to technological change that have now been in evidence since the late 1970s. These are:

1. The increasing speed and power of desktop computers which have now become extremely powerful and reliable machines which are commonly available for reasonably low outlay and,
2. The improvement in the telecommunications infrastructure, and related products, that allows the transmission of digital data at higher speeds with greater accuracy.

The convergence of these two aspects has been called the "Information Technology Revolution". This change in the method of communication has had many effects, some of which will be discussed later in the book. What has been especially noticeable are the changes that have occurred in everyday life as the introduction of computing and communication machinery has happened. Again many of these changes will be discussed later, but the various application areas that are the subject of later chapters have some interesting examples of profound change that has permeated society.

An earlier example is an interesting one. The bank cash machine is now ubiquitous in Europe and North America. These machines have led to a change in the working patterns of bank employees (ignoring the effect of redundancies), and the habits and customs of users and third parties such as retailers. It has also led to a change in the pattern of crime. Whereas, before the use of ATMs, the bank robber needed to rob the cashier, the ATM now provides a convenient point of entry!

In the world of work the use of computers is now widespread, and as a consequence the adoption and use of electronic trading or EDI is now becoming more common. Companies can concentrate on the core business functions without needing to be concerned with the information sent to customers and suppliers.

With the third aspect, that of multimedia, adding to this convergence there are now many possibilities that have before not been feasible. The use of information in multiple forms can enable a better understanding of the information that is communicated. It can also do the opposite if not used correctly. There is, therefore much more to be done to ensure that multimedia communication is as useful as it is portrayed and as beneficial to people as possible.

FURTHER READING

For a more complete treatment of computer communications the reader is referred to:

A. Sloane, *Computer Communications: Principles and Business Applications*, McGraw-Hill, 1994, ISBN 0-07-707822-5.

If you are interested in the historical use of different methods of communication over long distances a highly original book worth reading is:

G. J. Holzmann and B. Pehrson, *The Early History of Data Networks*, IEEE Computer Society Press, 1995, ISBN 0-8186-6782-6.

EXERCISES

1. The slide show and the picture album were mentioned in section 1.6. What other combinations of different media have been commonly used?

2. What are the codings used in the following communication examples

> A letter
> A video conference
> A Morse code message

3. Table 1.1 contains a list of factors governing the use of communication. Rank these factors in order, with a brief justification for each one for the following examples:

(i) Using a video conference facility in place of a 200 km car trip.
(ii) Using a video conference facility in place of a transatlantic flight.

2

COMPUTER COMMUNICATION

Summary: Communications, networks, applications and other communications topics used later in the book, with a special emphasis on multimedia needs of communication systems. Information quality and communication systems. Information transmission, errors and reliability

2.1 ELECTRONIC COMMUNICATION FUNDAMENTALS

In order to understand how multimedia communication is achieved and what the various problems are that affect its development, it is first necessary to understand the fundamentals behind any form of electronic communication. To do this requires some basic understanding of telecommunications and how computers are used to communicate over different distances, from a few metres to thousands of kilometres. Also a basic understanding of networking concepts is helpful. This chapter is designed to be a brief introduction to electronic communication and computer networks. Although it may be read as a self-contained text on the subject, it is almost impossible to put any great detail into so small a space. The reader is, therefore, directed to the various texts in the further reading list at the end of the chapter for a more in-depth treatment of these subjects.

In industry, commerce, education and the home, many people now use both computers and the telecommunications network as an everyday tool of their work. This has two effects. One is to increase awareness of what can be achieved by the convergence of the two technologies of telecommunications and computers, and the other is to breed a complacency about the capabilities of the systems, with users often not knowing if improvement is possible without major upgrades. This leads to a technological curtain being drawn between the user and the supplier of the systems, making users unaware of the potential of current systems and improvements that can be achieved with new ones or alterations to existing ones. Without a knowledge of fundamental principles it is difficult to judge whether a system will be an improvement or merely another alternative to the current *status quo*. This is the reason this fundamental knowledge is included here. It includes some basic results in computer communications and does so to allow the reader a more complete picture of the problems of communication in this environment.

The first aspect of electronic communication, of importance, is to relate the principles outlined in Chapter 1 to the particular environment of computer communications. Then the various aspects of each of the areas will be developed.

2.1.1 Digital and analogue communication

One area of confusion that often arises is the difference between analogue and digital communication. While the computer is a digital device which uses discrete signals to represent information, there are still systems in use in everyday life that are analogue, and in many respects analogue systems are more intuitively familiar. For example, turning a volume control on a radio or television allows a control of the volume of sound that is a smoothly varying analogue quantity. This feature of the analogue system is ingrained into everyday life, so much so, that digital systems tend to imitate it.

The real difference between analogue and digital signals is the number of levels that are used in the systems. In an analogue system there is a continuously variable state between the lowest and highest level. This can take any value in between these limits. In the digital system, there are only discrete levels. The digital system cannot have any other value but these. Figure 2.1 illustrates the difference on a single graph.

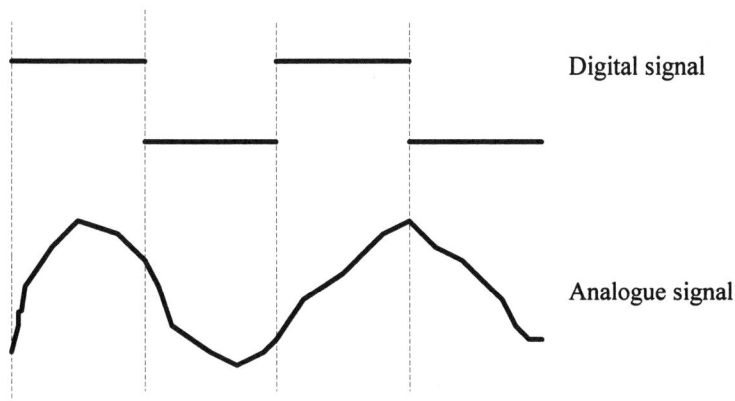

Figure 2.1 Analogue and digital signals

The upper part of Figure 2.1 shows the graph of a two-level digital quantity while the lower part shows a variable analogue quantity. The graph demonstrates the two extremes, the digital signal has two levels whereas the analogue signal has a varying signal that covers all the values between any two points. For the mathematically minded the difference is similar to the difference between integers and real numbers. The digital signal being the integer-like set of quantities.

What is now often done in digital systems is to use analogue to digital converters that can take an analogue signal and produce a digital output. To do this an analogue signal must be sampled at regular intervals to give an approximate digital version. This allows analogue devices such as sound sources, e.g. microphones, to interface with computers. An example of A-to-D conversion is shown in Figure 2.2.

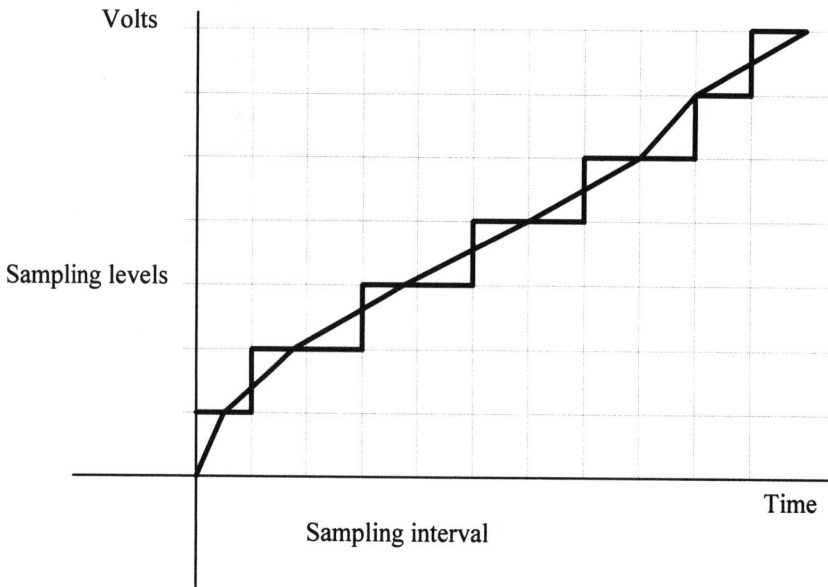

Figure 2.2 Digital sampling of an analogue wave

In Figure 2.2 the analogue signal has been approximated by a digital signal. In this example the approximation is not very close but in real systems the sampling frequency, which determines the interval at which samples are taken, and the number of sample levels can be controlled to give appropriate quality. For example, in sound sampling, the quality used on compact discs uses a sampling frequency of 44.1 kHz using 16 bit samples or 44100 samples per second using 65536 different signal levels. For speech only, this can be reduced to a sample rate of 8 kHz with 8 bit samples. The corresponding change in the quality of the signal is very noticeable.

In computers the information that is stored and processed is digital data. To communicate between two computers it is clearly best if a digital form of communication can be used as this will be the same for both of the communicating devices. Unfortunately, much of the communication that is carried out by computers is still done over an analogue system, so there needs to be some conversion of the information that is transferred to make it suitable for the medium of communication.

The traditional public switched telephone system (PSTN) is a system designed to be used by analogue voice signals. The restrictions imposed on the signals that can be transferred over the PSTN mean that it is not possible to directly communicate digital data. There needs to be a conversion between the digital output of a computer and the analogue PSTN. This is done by means of a technique called modulation and demodulation. The digital signal is converted to analogue by modulating it using an appropriate technique and the reverse is done at the other end of the communication link where the digital information is recovered. The system is illustrated in Figure 2.3.

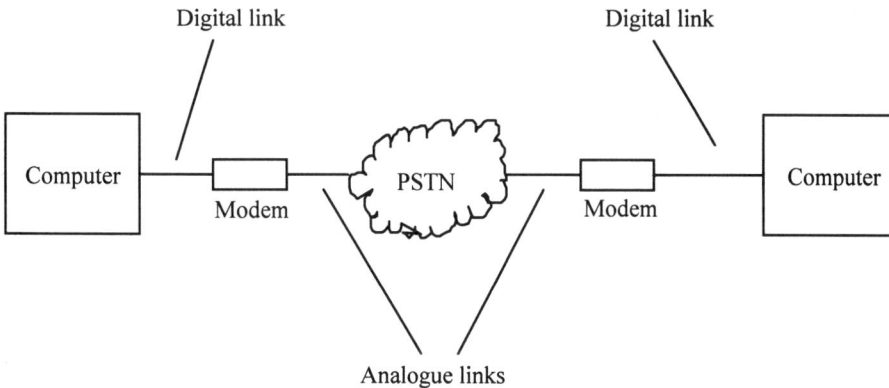

Figure 2.3 Using the PSTN for computer communication

This is a good demonstration of the communication principles outlined in Chapter 1. To use a channel (the PSTN) the signal needs coding in an appropriate way (modulation). This encoded signal is then transmittable over the PSTN and can be decoded to produce the original digital signal at the other end. The effect of any errors at this stage being ignored. There are, however, techniques available for error control which will be outlined in section 2.1.3.

To be able to use any channel it is essential to know whether it requires analogue or digital signals. The most appropriate channels for computer and multimedia communications are, undoubtedly digital channels. Unfortunately in many situations the use of analogue channels is required. It is, therefore, helpful to be aware of the limits of using either of these types of channel. These limits are discussed next.

2.1.2 Capacity, bandwidth and multiplexing

One minor point of confusion that can arise is the difference between data that is transmitted in parallel and serial. The majority of data is transmitted one bit at a time, i.e. in serial, but there are some systems on computers, e.g. most printer ports, that use parallel transmission. Parallel transmission uses a number of separate channels to transmit a number of bits simultaneously, in parallel. The channels considered throughout the rest of this book will be serial unless stated otherwise.

The usual measure of a digital serial channel is its capacity. This is normally given in bits per second. It is, therefore, a measure of how much data can be transmitted by the channel in a second. The rate of a particular channel will depend on the type of channel it is, i.e. what the transmission medium is and what equipment is being used. This will inevitably be some standard type of access device which will be set to access the medium at pre-determined data rates.

There are also problems in calculating *effective* data rates when a channel is shared between different sub-channels and users. Any other use being made of the channel will have a direct bearing on the data rate of a particular communication. Channel sharing is commonly used in situations where there is only one medium between different communication points and the total capacity is not needed for a single channel. The methods used for sharing are generally termed multiplexing. The general idea is shown in Figure 2.4.

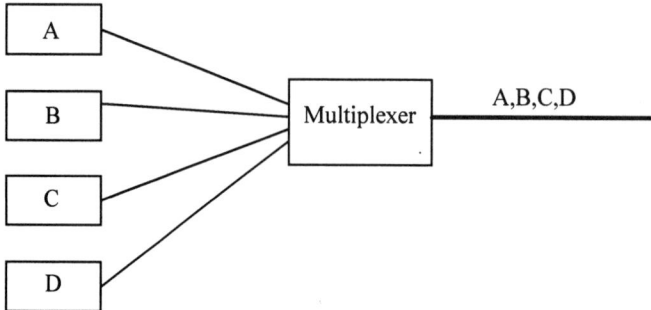

Figure 2.4 Multiplexing

Figure 2.4 shows how four computer communications channels from computers A,B,C and D can be joined together by a multiplexer and form a single combined channel carrying A,B,C and D simultaneously.

There are a number of means of multiplexing signals. The two most common methods are known as time division multiplexing (TDM) and frequency division multiplexing (FDM). These both split the available channel into smaller sub-channels for the use of a single communication link.

TDM uses the principle of dividing the channel into slots and allowing different sub-channels to have alternate use of the slots, either on a regular rotation or by a demand-based system. FDM divides the available channel into sub-channels of lower capacity by modulating the signals at different frequencies within the band used by the medium. These are illustrated in Figure 2.5.

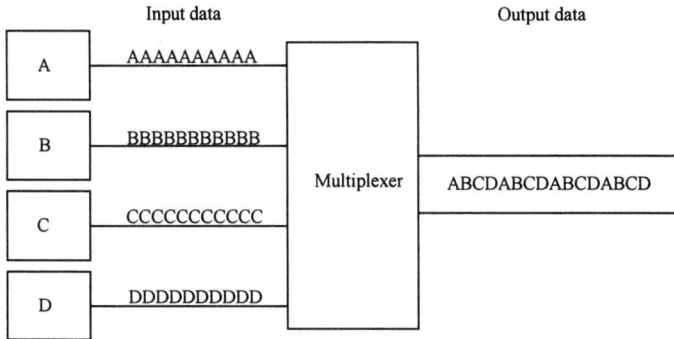

Input data   Output data

A	AAAAAAAAA
B	BBBBBBBBBB
	Multiplexer	ABCDABCDABCDABCD
C	CCCCCCCCCC
D	DDDDDDDDD

Figure 2.5 Time division multiplexing

The TDM used in Figure 2.5 shows four channels being multiplexed into one combined channel. In this case, if the four channels each have a capacity of 64 kbps the combined channel will need to have a capacity of 256 kbps. The theory for FDM is left until later.

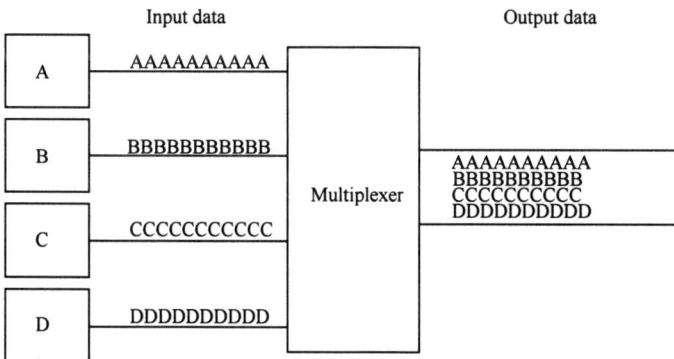

Input data   Output data

A	AAAAAAAAA
B	BBBBBBBBBB
	Multiplexer	AAAAAAAAA BBBBBBBBBB CCCCCCCCCC DDDDDDDDD
C	CCCCCCCCCC
D	DDDDDDDDD

Figure 2.6 Frequency division multiplexing

There are other methods of multiplexing a channel, but all will have the same effect. The true capacity of the combined channel will be divided between the different users. In the example, four sub-channels share the main channel's capacity.

In an analogue system capacity can be calculated, within limits, by using the Shannon-Hartley law. This relates capacity to bandwidth and signal to noise ratio which are measurable analogue parameters of a channel. The Shannon-Hartley law states that

$$C = w \log_2(1 + S/N)$$

where C is the capacity of the channel, w is the bandwidth and S/N is the signal to noise ratio.

Bandwidth is the range of frequencies that can use the channel. The signal to noise ratio is the ratio of user signal to background noise, although some restrictions apply to noise that is included in this calculation. S/N is normally quoted in decibels (dB).

A rough calculation shows that for a channel of 5 kHz bandwidth and S/N of 20 dB the channel capacity appears to be about 33 kbps. This calculation does not take account of other noise types and can be effectively reduced in practice. It does, however, give an upper limit on capacity that will not be increased by improving communication techniques, only by an improved channel. For further details of the method of calculation using the Shannon-Hartley law the reader is referred to the author's previous book, details of which are in the further reading section at the end of this chapter.

One final point that is now a cause for confusion is the use of the term bandwidth. Although it has been defined here to mean a frequency range in an analogue channel it is frequently used to mean capacity. This is an acceptable misunderstanding if the Shannon-Hartley law is considered, as the bandwidth and capacity are linearly related. It is therefore usual now to refer to bandwidth requirements when the real quantity needed is capacity.

2.1.3 Protocols and error control

To communicate effectively over a channel requires that the communication has some structure. This is as true of everyday communication as it is of computer communication. The communicating devices need to establish contact, ensure a reliable flow of information between each other and then release the channel after the information transfer has been completed. To do this requires a protocol. Protocols provide the required structure to communications. The principles of protocol design are listed in Table 2.1.

A brief description of each of these aspects of protocol design follows:

The **message format** refers to the logical structure of each of the messages exchanged by the two communicating devices. This allows items such as an error control code or a start of text character to be placed in a specified location within the message.

Table 2.1 Principles of protocol design

Message format
Error control
Acknowledgements
Time-outs and re-tries
Sequence control
Flow control
Recovery
Data transparency

The **error control** mechanism can have two functions. The simplest is to detect when errors have occurred and allow the protocol to deal with the resulting problem by repeating faulty information, or there can be an error correction mechanism built in to the message that allows correction on receipt. This second option does, however, need more data to be sent and the method used requires a trade-off between efficiency and the need to correct errors. In situations where low error rates occur, detection is usually adequate. For noisy environments error correction techniques can prove worth while.

An **acknowledgement** scheme is required so that there is a specified response to incoming data and the transmitter then receives confirmation of delivery or otherwise.

If data is lost or a communication device goes out of service a **time-out and retry** mechanism is usually used to re-establish contact. The number of retries is limited to prevent locked channels and the time-out is calculated to allow for the communication speed and response expected on the channel.

A **sequence control** mechanism is used to control the order of data being transmitted. Messages are often sent in the order they occur in a communication, but it is possible for them to arrive out of sequence and a sequence control scheme allows for re-ordering of messages to reform the original information.

Flow control is needed when the communicating devices have different speeds. If a fast device needs to communicate with a slower one then there needs to be a mechanism to stop the communication when the slow device has enough data to process, and to restart later. The reverse procedure is also used.

Recovery is tied to other aspects already covered. To be able to recover from a communications problem the protocol needs to be designed with this in mind. So sequence control, time-outs, retries and acknowledgements are all important here.

Finally, a protocol needs to be able to handle all types of data. If it does it is said to have **data transparency.** That is, it can carry all combinations of binary digits, not just ASCII character data. This is a particularly important characteristic needed for multimedia communication.

All these aspects need to be addressed if a protocol is to have a workable specification. Error control is especially important as faults in data can have profound effects.

If the sequence

3124.00

is transmitted and a one bit error occurred the received data might be either

1124.00 or 7124.00

If these refer to deposits into a bank account, the first is a loss of 2000 and the second a gain of 4000, in whatever currency is used. Errors of this type could break the bank or the depositor!

2.1.4 Information quality and reliability

As the example at the end of the last section demonstrated the reliability of information can have far-reaching effects. While this sort of effect is not common, if adequate safeguards are not incorporated into a communication scheme errors of this type can occur. If transmitted information is to be reliable the communication process needs to incorporate features that are required by the rules of communication as outlined in Chapter 1, and to use the protocol rules outlined in section 2.1.3. With these rules a communication can be completed that allows data to be transferred reliably. It should also enable the original information to be recovered from the data, this will depend on the coding technique used and whether it is appropriate to the needs of the communication. Usually it is enough to assume that the reliability of data will ensure the reliability of information.

If the quality of the information is considered then the communication process only plays a partial role as long as the communication is reliable. The quality of information is more likely to be affected by choices made before the communication takes place. For example the following features are relevant to the delivery of information from an information provider and are equally relevant to any communication of information, in electronic form or otherwise.

Timeliness of the delivered information
Speed or frequency of the information provision
Completeness of the information
Selectivity of the information
Relevance or specificity of the information

These qualities are all required, if relevant, for any communicated information to have the necessary quality. In terms of electronic communication of information, the first two items, timeliness and speed/frequency may be affected by the communication infrastructure, but are just as likely to be affected at the input stage of the process. For example, if a computer terminal is being updated with share prices, it is essential for dealers to have accurate timely information. If the input takes too long the data will be out-of-date before it is used and the whole system would be worthless. The other features, completeness, selectivity and specificity are all related to the information input into the process, not the communication process itself. To allow easy use of information, these aspects should be addressed outside the communication process.

There are, however, times when these aspects can affect the use of communication systems for information retrieval. A particular case is the use of the Internet to browse information sources. Because of the mainly unstructured approach taken by Internet users and information providers as a whole these three aspects of information *quality* are low in this environment. There is, however, a vast *quantity* of information.

2.1.5 Compression

One aspect of communication and information transfer that affects any system designed to transfer large quantities of data is the use of compression techniques. In Chapter 1 the communication principles included mention of redundancy as a fundamental requirement to ensure that errors could be controlled. The use of compression techniques allows redundancy in input information to be minimised so that the amount of data transmitted is smaller. The communication process may well add data to ensure error-free communication (as far as possible), but this would then be within the parameters of the protocol used to transfer the data. The use of compression reduces redundancy in data before the communication process. For example, if a text document contains a lot of repeated "whitespace" characters (spaces, tabs, etc.) then it can be compressed into a shorter form, This may be done by using some form of run-length encoding which effectively replaces long runs of the same character by a character +, a count variable or, alternatively, by a code word which indicates the length of data compressed. This latter technique is used in the encoding of bits in fax transmission.

Many forms of information are highly redundant. This allows considerable degrees of compression to be applied and sometimes compression results in a slight loss of the quality of an image or a similar data object. If this is the case the technique would be termed *lossy* whereas, if the original data is fully retrievable, it would be termed *lossless*. When large data objects are compressed the communication process can work more efficiently, having less data to transmit. This, in turn, improves some of the other information quality parameters (e.g. timeliness) and allows a broader range of information to be incorporated in lower capacity channels. Some compression techniques will be described in Chapter 3 for particular media types.

2.1.6 Standards

Communication standards have been discussed earlier. Their importance has shaped the communications infrastructure of the modern world. The fact that telephone systems around the world can inter-work and exchange calls is due to the widespread adoption of communications standards by the telecommunications companies and service providers. The same is not always the case in other industries, with the computer industry being an example of standardisation after the event. However, there are many standards in both communications and multimedia which will be described and discussed in this and later chapters.

For the communications industry there are now a number of choices for how the technology will develop over the next few years. These choices depend on the prevalence of competing technologies and systems which are still in their infancy. However, there are some trends emerging which are discussed in section 2.4. In communication, standards have always had a pivotal role in technological development, with success of new technology being dependent on its acceptance as standard by the big players. There is now a similar situation occurring in higher speed data networks, with various standards being proposed and some of them adopted by the relevant bodies. This inevitably helps the communications industry, by allowing them to concentrate production and development on standard product lines, and also the user, as they will know a standard product will be compatible.

Without standards individual communication products would have to compete against a whole range of others of similar, but different incompatible characteristics, and the possibility of being able to communicate widely with a single device would be a remote possibility. In multimedia communication this standardisation process is at an early stage and has not yet borne much fruit, although there are the early signs of some useful developments in areas such as video conferencing and related activities.

2.2 COMMUNICATION SYSTEMS

To understand the complex area of communication systems there are a certain number of basic types of medium that need to be understood along with the various characteristics that they display. This will, in turn, allow some degree of qualitative assessment of the technology of communication and any advance that is made.

2.2.1 Technology for communication

The ability to communicate has been shown to derive from a number of principles which govern all communication, including the electronic version. Therefore, to communicate effectively using computers and communications technology a channel must be available, the equipment used must be able to utilise the channel, and a technique must be used that is appropriate to that channel allowing error control and incorporating redundancy techniques. There are, however, considerations from the opposite end of the problem. If the data that is to be communicated is particularly massive, e.g. high quality sound or video, and needs to be done in a particular time so that real-time use is made of the information then there are factors pulling the choice of the technology.

As an example, if the only channel available is a PSTN connection with a capacity of 28.8 kbps it would be impossible to send uncompressed video data along the channel in real time. If the recipient was prepared to wait for the image, i.e. a playback system is used, then a small amount of video could be sent, but even then a few frames would take a long time depending on the frame size, refresh rate, etc.

So to have a quantitative grasp of communications channels the medium used to carry the data must be known, or at least the capacity of the channel. (The capacity can usually be inferred or calculated from the knowledge of the medium and its characteristics.) It is not always a matter of choice. The situations used in the later chapters will all be familiar to the reader, and what is commonly a problem is the fact that infrastructure improvements in many of these situations (e.g. education or the home) are not always a cost-effective proposition. So, many users of communication channels need to use existing channels that are available, and use them in an effective and efficient manner to capitalise on capacity that is available. This may not be the case for much longer, as capacity in digital systems becomes cheaper and the installation of new media becomes more widespread (e.g. cable TV in the home and elsewhere).

2.2.2 Transmission media

There are a number of transmission media that have been used for digital communication systems, the way they are used may vary between application and the distance they are used, but the general characteristics are still relevant.

The primary medium that has been in use for many years, and still is in many places, is cable. This comes in many forms and at different transmission quality levels, but is generally of low to medium capacity.

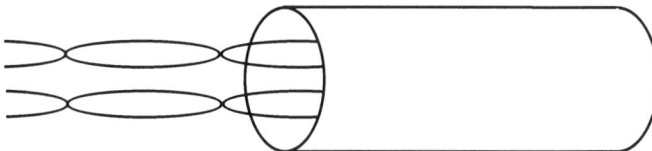

Figure 2.7 Twisted pair

There are two main cable types in use for computer communication purposes. Both use electrical signals to communicate. The first is **twisted pair** as illustrated in Figure 2.7. This usually comes in multiple pair cable with an insulator on the outside and around each cable. It can be either **unshielded,** or **shielded.** The conductor pair is twisted to decrease noise and interference and the shielding is incorporated to further reduce electromagnetic noise effects. Consequently, the best twisted pair cable is shielded (called **STP**), but it is more expensive, and the lower quality is unshielded (**UTP**). There are, however, different grades of UTP that are used in data communications applications. These have guaranteed quality levels and can be found in most data communications suppliers catalogues. Common specifications are level 3, 4 and 5 in increasing quality order. Many installations now specify a minimum quality standard for cable installation. Better quality cable allows greater capacity, or longer distance communication.

The second cable type used is **coaxial** cable as illustrated in Figure 2.8. This is a conductor surrounded by another conductive shield, the two being separated by insulation, and surrounded by further insulation on the outside. Again, there are different qualities and types of coaxial cable and the correct type needs to be used to guarantee the data rate required.

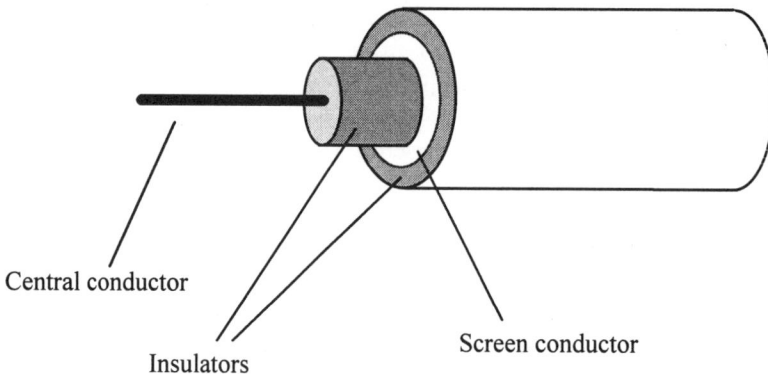

Central conductor

Insulators

Screen conductor

Figure 2.8 Coaxial cable

A more modern type of data carrier is **fibre optic** cable. Commonly called just **fibre,** which uses light to carry data instead of the electrical signals used in twisted pair and coaxial cables. This is more difficult to install and has generally been more expensive but as new techniques for handling fibre are developed and used the price has fallen and it is now relatively cheap to install if the price per Mbps is calculated. Fibre is illustrated in Figure 2.9.

Again, there are many types of fibre available and the exact type required for a particular application would depend on many factors, including the standard system being used, the actual location and installation environment of the cable, and the data

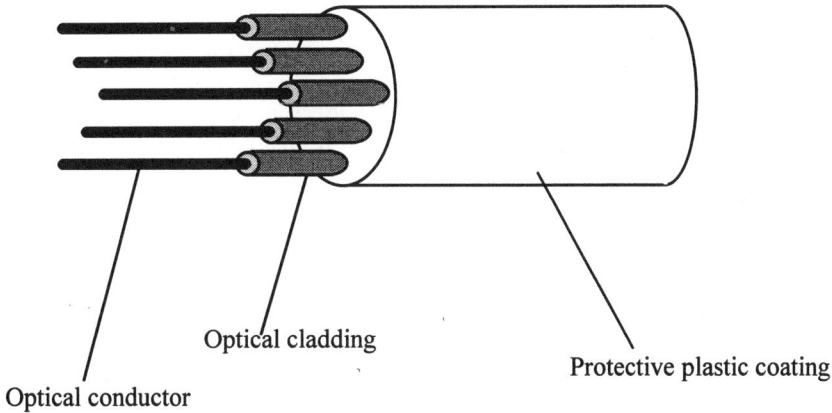

Optical cladding

Protective plastic coating

Optical conductor

Figure 2.9 Optic fibre

rate required. However, in general terms fibre is capable of data rates many times higher than cable, over greater distances.

Finally, one other medium that is often used in both local and long distance applications is radio. Here the signal is modulated with a radio carrier and demodulated on reception. In the local environment a low power transmitter is used to confine the signal to the area of interest e.g. an office building or local campus, in the long distance example it is often a point-to-point link that is used to connect transmitter and receiver with low power directional radio equipment.

The other main area of radio use is in cellular systems. These require complex base station configurations to give coverage over wide areas and have primarily been designed for voice communication, but the newer digital systems are equally capable of data communication over the same systems, allowing mobile data communications to be used in addition to the more traditional land based systems. The characteristics of radio and in particular mobile radio usually only allow low capacity signals as bandwidth is limited.

2.2.3 Characteristics of communication systems

The main factor of choice for any communication system is the capacity it offers the user. The application that is to be used and the facilities expected from it will determine the capacity required. So if two users wish to communicate with video and sound in real time then the system will need the capacity to transfer the amount of digital information in these two media types. This will require a high capacity channel or in some circumstances a number of aggregated low capacity channels working together.

Whatever system is used the various characteristics of the system will need to be taken into account when deciding on the protocol and error control system to be used. The nature of the noise affecting a channel will determine the error control mechanism used. For example, in a fibre optic channel (a low noise environment) a simple error detection scheme is usually more than adequate, yet in a cellular radio environment with variable signal strength and hand-off between cells, the error control system needs to be able to cope with error bursts and other problems without too many repeats of data.

2.2.4 Example systems

There are a number of communication systems in common use that are worth further examination. The commonest system, and most universal of all the possible technologies, for data communications, is the PSTN. This was originally designed for voice communication and still allows analogue voice to be used between two users, although a large number of systems are actually digital. This original design restriction and the fact that systems need to inter-work has meant that the available capacity for data traffic is fairly low. The current (1996) fastest standard for high speed modems on the PSTN is 28.8 kbps. This is adequate for many data applications but not very usable for large data objects such as video clips or high quality sound.

The fully digital version of the PSTN is the integrated services digital network (ISDN). This comes in a number of versions based on the basic channel (B-channel) of 64 kbps. The standard offering is 2B + D. This is 2B channels plus a 16 kbps D-type signalling channel. This is provided over copper cable. The B channels can be aggregated to form a 128 kbps channel. Another common option is the 30B + D combination which has 30 64 kbps channels which allow a total data rate of nearly 2 Mbps. This is provided using fibre optic cable. Depending on the channels available, most data types can be transmitted over ISDN, in fact, a number of video conferencing facilities have been demonstrated using ISDN and high quality sound transmission has enabled radio stations to use programme presenters working from home over ISDN.

Cellular radio allows mobility, but the price to be paid is in the capacity of the data communication channels available. With a limited bandwidth and a variable signal to noise ratio the best, reliable capacity available is about 9600 bps. This limits communication to the most basic data types.

Probably the biggest development in communication and the availability of increased capacity is the development of ATM (asynchronous transfer mode). This is a system that communicates using small cells of data and can transfer them around a switched network at very high speeds. The importance of ATM lies in its ability to offer a range of capacities to suit different applications from low data rates to the high speeds required by complex multimedia communication applications.

2.3 NETWORKS

The connection of computers to enable communication has inevitably led to the creation and invention of complex computer networks to inter-link many computers simultaneously. This enables exchange of information between any two computers on the network, and sometimes allows broadcast facilities so that any user can communicate simultaneously with all other users. It also establishes new methods of operating for many users who can access information that is decentralised or distributed around the network with common files available to all users or selected groups with common access permissions.

2.3.1 LANs and WANs

Computer networks are classified according to their area of operation. So a network that belongs to one organisation is called a **local area network** (**LAN**) and one that operates within a larger area, such as country is called a **wide area network** (**WAN**). This is, however, a great simplification since often a LAN can be used over very large areas by the introduction of remote bridging techniques and wide area systems could be used in relatively small domains. So there is often a secondary definition based on the technology and the ownership of it. Namely, that LANs use access methods and technology that is suitable to higher speed networks in localised environments and WANs use technology that is generally available to users via a service provider, such as a PSTN.

This is necessarily a little vague, but the differences are commonly understood. For example, a CSMA/CD bus network is a LAN technology and runs at speeds that dictate use in small areas, by virtue of cable length limits and so on. X.25 is a wide area access protocol that works over the type of line that is available between widely dispersed sites. LANs and WANs have different characteristics and these should become clear on reading the following sections.

2.3.2 LANs

Local area networks are generally of three different topologies. A topology being the basic layout of the interconnections in the network. These three types are. *star, ring* and *bus* networks. These are all illustrated in Figure 2.10.

Fundamentally, most local area networks are of the bus topology as this has proved to be the best seller over many years. Star networks have been used in the past to connect terminals to mainframes and have the advantages and disadvantages of centralisation, but it is now common to find what appears to be a star network which is in reality a bus network as illustrated in Figure 2.11.

In Figure 2.11 the computers are connected by cables to a hub in a star configuration but this functions as a bus network as the internal connections are as the bus-type network. This topology has become more popular in recent years as the ease of connection and maintenance is greater than with conventional bus networks since all the connections are concentrated in one place.

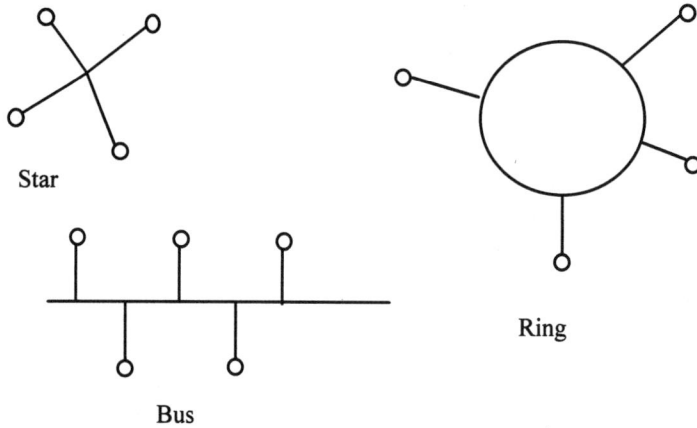

Figure 2.10 LAN topologies

Ring topology LANs are now less common than bus types, and have generally been the least popular of the two types. They do, however, have certain advantages in use that are related to their ability to avoid congestion. Bus networks generally use a contention mechanism whereby capacity and transmission slots are used by a contention mechanism, i.e. each station that needs to transmit does so when the bus is free. This can cause problems when all stations are heavily loaded. In ring networks, the contention resolution is by token passing and if a station has control of the token it can use the network. The ring scheme is much better under heavy load.

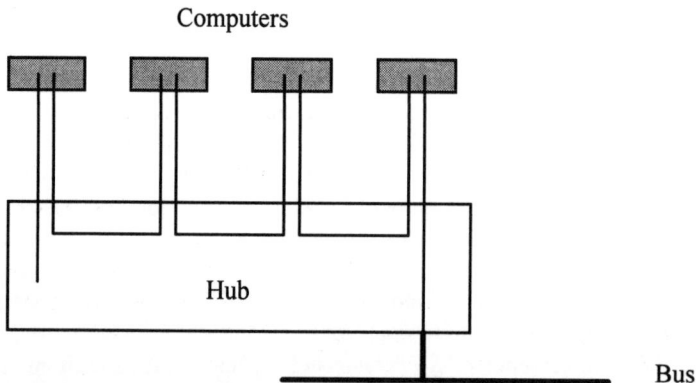

Figure 2.11 Star-bus network

Unfortunately, the speed of operation of LANs is not sufficiently high to allow much multimedia communication to be used without the user having total control over the network medium. As both of the schemes described have been designed to share the available capacity between users, new schemes are required if multimedia is to be a serious application on LANs.

Generally, the standard designated capacity on a common CSMA/CD bus network is 10 Mbps and on a token ring network it is 4 Mbps or 16 Mbps. These capacities are shared between the complete user base of the network, so the more users that use the network, the slower it will be although there will be occasions when a user is able to use most of the capacity, these will not be guaranteed at any time

2.3.3 High speed LANs

To alleviate the bottlenecks of current standard LANs, manufacturers have put forward newer proposals to the various standards bodies. This has resulted in the latest LAN standard at 100 Mbps. There is now a range of equipment available at this higher speed which is consequently more expensive and less widely available than the standard 10 Mbps equipment that has been the market leader for a number of years.

The increase in raw speed will go some way to increasing an individual's access to the capacity available but is likely to allow networks to incorporate more users and so limit the share of capacity for individual users again. To gain an increase in available speed the technology needs to be changed to allow users access to the full capacity of the medium being used. This is because the protocol used for access in CSMA/CD networks is a fundamental weakness in the design of LANs and is the main cause of reduced utilisation in multiple access media. So, the higher speeds may allow some improvement in the use of multimedia applications it does not guarantee data rate.

Physically, higher speed can usually be achieved by closer specification of cabling and equipment allowing the data rates to be increased on similar cable to the that used on lower speed LANs.

2.3.4 Switched LANs

A more effective change to the specification of LANs is to replace hubs with switches. These are based around fast internal busses that can handle traffic at much higher rates than the individual lines into the switch and so provide a switched facility that allows users full capacity connections.

LAN switches also allow variable speed connections to be used. For example a set of user computers could be connected at 10 Mbps and a server could have a 100 Mbps connection. Uniform speed connections are, however, more common at present. A typical switch would be as shown in Figure 2.12.

It is also possible that the high speed connection is used to connect together a number of switches into a more complex network serving many users. Typically, each switch covers between 6 and 24 ports, or connections. So, to cover even a medium sized enterprise the number of switches needed will be a substantial part of the cost of the network.

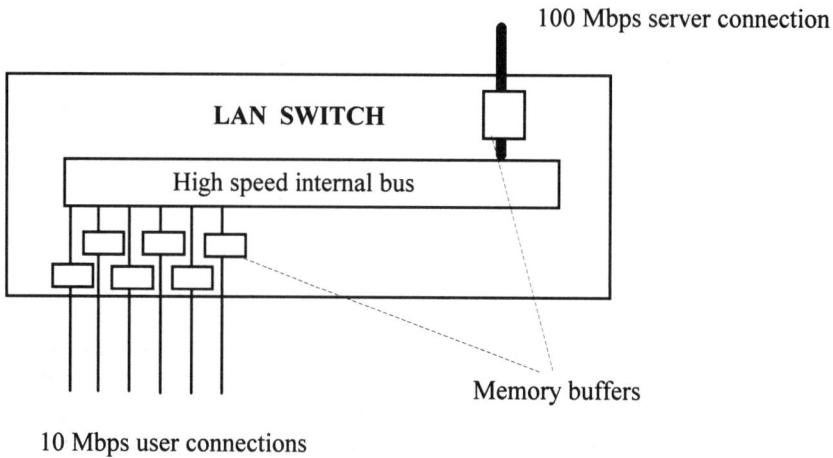

100 Mbps server connection

LAN SWITCH

High speed internal bus

Memory buffers

10 Mbps user connections

Figure 2.12 LAN switch

The advantage of LAN switching, however, is that the capacity available to the user is better guaranteed than in a contention system. Where as the traditional CSMA/CD network relies on usage being shared between the users by virtue of their wanting to use the network at different times, switching cures the bottleneck by supporting high data rates inside the switches allowing users full use of the capacity of their LAN connection. This gives users a better capacity for multimedia applications and allows some guarantee of bit rates to be given for higher speed use. Since the actual data rate on each LAN segment is within the slower speed specification the cabling required can be of similar quality to standard 10 Mbps systems.

2.3.5 LAN standards

Local area networks have been designed and invented by a number of different manufacturers and mostly the results of this diversity has been the distinction between the various systems that have emerged. To make LANs a worthwhile investment for users a degree of standardisation needs to be employed. This task has been conducted mainly by the IEEE, the Institute of Electrical and Electronic Engineers in the United States. Their work also feeds into the international standards process of the ISO (International Organisation for Standardisation). The various LAN standards that have been formulated are known as the IEEE 802 standards. These cover most types of LAN and are under constant revision. For example, IEEE 802.3 is the basic CSMA/CD bus LAN standard and IEEE 802.11 is the wireless LAN standard.

2.3.6 WANs

Wide area networks are very different from LANs. The speeds of operation available to users are generally much lower than in a LAN environment and the use of shared media outside the users control does not allow guarantees on data rates to be given. Generally WAN connections are provided by public bodies or service providers such as telephone companies, and a guaranteed *access* speed is provided. The network itself is transparent to the user and this allows the provider to structure it, internally, in any way that is appropriate to the delivery of services. However, there are standards applicable to WAN access and these are fairly commonly available. An example of a more common standard is X.25, and of the newer standards, Frame Relay and ATM are becoming more popular.

Mostly WANs use a system called packet switching that sends packets of information between nodes on the network. These then effectively multiplex communication down the various links of the network sharing the available data rate between the users. Network links are often leased lines that are permanent connections between nodes on a network with a predetermined capacity. This capacity then varies with the traffic rate from the users for each link. Consequently, a knowledge of the use of a network is needed to plan the need for capacity in each node-to-node link. Some of the traffic rates will also be determined by the algorithm used for routing packets through the network.

2.3.7 Narrow band restrictions

Most wide area provision is relatively low capacity and effective data rates are variable. This is caused by the multiple use of links in the network by many users communications. For example, a WAN may be configured as in Figure 2.13. Here the node at D is a bottleneck in the system if the traffic is mainly end to end, i.e. A to G or A to H. But, if most traffic is confined to either the left side or the right side of the network there is little point in adding redundancy if little use is made of it.

All the communications from A to E, F, G or H will need to pass through D, but there are two routes to D from A. Either A to B to D, or A to B to C to D. Spreading the traffic between these two routes will enable users to have better response and, therefore, better effective capacity. So although individual links may have a given capacity, there are a number of factors that determine the usable effective capacity. These will include the topology and design of the network, the capacity of individual links, the routing algorithm used and the total load on the system at any one time. Together these factors place a restriction on capacity that is variable and unpredictable

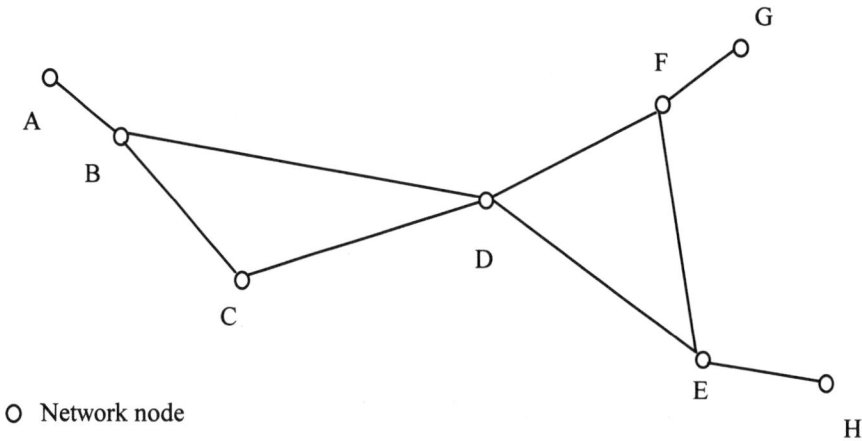

O Network node

Figure 2.13 A typical WAN

to the individual user. Increasing link capacity will go some way to improving the situation but the tendency is for users to use whatever capacity is available and if this happens the effective data rate will again decrease for users. So a new method of working is required. This is provided by using a different switching technique.

2.3.8 Broadband systems

The future of communication in WANs appears to lie with asynchronous transfer mode (ATM) and broadband ISDN (B-ISDN). These two developments go together in the WAN area. At the physical level, B-ISDN is a digital service of high capacity to allow multi-megabit data rates. On top of this goes ATM which is a fast cell-relay system designed to work at a variety of high speeds. A standard speed of 155 Mbps being chosen at present with future upgrades to 622 Mbps and beyond being foreseen.

ATM uses small cells and high speed switches to set up virtual circuits between end-points; this gives a connection between the two ends of a communication and allows a guaranteed capacity for each communication. ATM does not set up connections if the capacity cannot be guaranteed. The cell size is 53 bytes, which consists of a 5 byte header containing virtual path and virtual channel identifiers and a set of flags, and a 48 byte data field. This is shown in Figure 2.14.

ATM is a connection-oriented service. The path of the communication is set up when the communication is started and cleared down when it is over, although semi-permanent connections can be allocated for use, such as in remote connections between LAN segments. In fact, one of the strengths of ATM is its ability to link LAN level and WAN level communications. ATM can be installed as a local area option and used to link local high speed communications to the wide area.

Virtual path id	Virtual channel id	Flags	DATA

———————— Header - 5 bytes ————————   Data - 48 bytes

Figure 2.14 ATM cell structure

By virtue of its ability to guarantee data rates ATM is seen as the only real option for future multimedia communication requirements in both the local and wide area. However, ATM is still an immature technology and products are not sufficiently available to enable the switch from older technologies easily. There will, therefore be a number of years in which the previously discussed technologies will be used for a variety of multimedia communication tasks and the sub-optimal performance of the networks will need to be part of the equation when using the systems.

2.3.9 WAN standards

Currently the most prolific standards that are used in WANs are the Internet suite of protocols, commonly called TCP/IP after the transmission control and internetworking protocols. These are a set of related standards that cover a layered protocol architecture from application down to the interface with individual sub-networks. The international standard for open systems interconnection (OSI) is also an important model in wide area connections. To a certain extent the two architectures fit together since the sub-network layer in the Internet protocols is not defined, but often relies on the OSI model to provide the basic structure. The protocol layers in each architecture are shown in Figure 2.15.

Most networks have been described using the OSI model. LANs and WANs have both fitted into the lower three layers used for the description of protocols within a single network. At higher levels the Internet internetworking protocol is now used extensively to interconnect networks over any area, and is commonly used to exchange data in the LAN environment also. At the higher end the applications that are used on WANs and LANs come from both models. In commercial environments OSI protocols such as X.400 are commonly used to exchange messages and EDI data, whereas in less commercially sensitive areas the Internet protocols, such as FTP (file transfer), SMTP (message transfer) and HTTP (hypertext transfer) are in widespread use.

Access to WAN services is, in effect, governed by the ability of the user to provide appropriate communication in the standard protocol used by the network they wish to access. So if the Internet is the target the users needs to have access to a TCP/IP package that will link into the network. This type of software is now readily available and becoming easier to install as systems become geared towards network access and the use of connections to networks.

When multimedia is considered in addition the prospects alter, since the rapid growth of multimedia has caused a glut of media standards to appear on the networks. It is only with time and use that a comprehensive, universal standard for various media types, combinations and other aspects will stabilise.

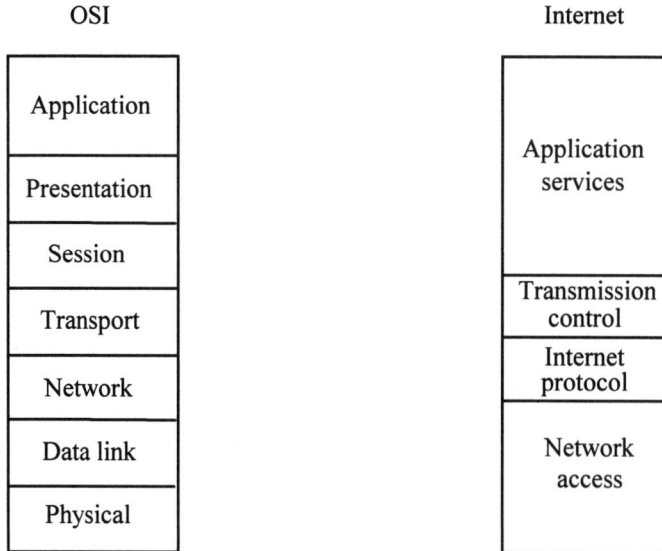

OSI   Internet

Application
Presentation
Session
Transport
Network
Data link
Physical

Application services
Transmission control
Internet protocol
Network access

Figure 2.15 OSI and Internet architectures

2.4 TRENDS IN COMMUNICATION TECHNOLOGY

There are a number of discernible trends in communication that will have an impact over the future direction of networks, communications and multimedia.

The Internet is now a powerful force in computer communication. It has enabled many users to communicate who would not have done so otherwise. It will be discussed in more detail in the next section.

The role of standards cannot be dismissed as the maturity and implementation of standards allows users to create networks that will communicate with others. Standards will become increasingly important in the search for communication systems that can handle multimedia in a way that is as easy for users as possible, without the need for complex conversions. They will also be increasingly important in media and document definition and other areas that impinge upon multimedia communication.

Finally, there are a number of developments in communications technology that are being made that will have some impact on the various environments that are of interest later in the book, namely the home, in industry and education. Amongst these developments are the definition of fibre channel, a standard aimed at even higher bit rates than ATM. It is defined for bit rates up to 1 Gbps over various transmission media from STP and Coax to different optical fibre types. The bit rates being defined for suitable distances for each different medium.

Crucial to the take-up of these technologies will be the ability of users to access them. This will be determined to a large extent by the infrastructure developments that take place between various countries to provide suitable connections to homes, industry, commerce, educational establishments, hospitals and other interested user communities. The quality of the infrastructure will in turn affect the extent of the multimedia communication developments in each country and the communications between countries.

2.5 THE INTERNET

As the largest collection of connected computers in the world the Internet forms an important communication tool. As such, a brief description of how the Internet works and what it can do is appropriate. It does not form the only network in the world and is unlikely to do so as there are many transactions that take place on private networks and on publicly available value-added networks that are more appropriate there than on the publicly open Internet.

2.5.1 Overview

The growth of the Internet has been phenomenal. From the 1970s until the beginning of the 1990s the Internet was a collection of mainly US networks that connected together various universities and research establishments to form a network for sharing knowledge, mainly for research purposes. The interconnection of other networks world-wide and the increased availability has given the Internet a world-wide group of users of many millions and it continues to grow. However, it is clear that substantial redesign and restructuring needs to take place to make the Internet able to cope with this phenomenal growth and the user's desire for more access to multimedia.

The Internet uses TCP/IP to communicate packets of information between end user computers that are connected to it. The connections can take the form of permanent connections through high data rate links or, for individual users, dial-up access via telephone or ISDN lines can be used. On permanent connections an IP service would transport data packets between applications based on the underlying network access protocol. On dial-up lines an underlying protocol is needed to allow TCP/IP packets to be sent. This can be provided by either SLIP (serial line IP) or PPP (point-to-point protocol). Most users connecting a single machine to the Internet on an intermittent basis would use this type of dial-up connection.

TCP provides connections between end points in the communication process, so it can be used for reliable transfer of information between applications and users. There is also a less robust protocol available called UDP (universal datagram protocol) which is used when reliability is not required. For instance, live video can use UDP as loss of data packets will only degrade the image not make it unusable (unless a high proportion are lost). These protocols form the basis of the transport mechanism in the Internet, above them are the various application service protocols that are used.

2.5.2 Current use and trends

The major applications available on the Internet are electronic mail, file transfer, news groups and hypertext transfer known as the World Wide Web (WWW). Until the advent of the hypertext protocol, HTTP, the main traffic was in email and file transfer and this consisted of mainly small data transfers which only used a limited capacity, and the network was able to move this mainly time-insensitive data easily. Since HTTP has become common across the network, two factors have changed the nature of the data transferred. Firstly, more large data objects have been transferred putting the network under increasing strain. Secondly users have become more interactive in their use of the network. Whereas email and file transfer were off-line background activities, Web browsing using HTTP is designed to be an on-line activity.

This trend is set to continue, with more media types such as quality sound and video becoming common in hypertext documents and users becoming more time-sensitive in their use of interactive applications such as browsers. Since the origins of the Internet were as an email and file transfer network the increasing strain is bound to show in the performance and response times of data transfers, especially where long distances and limited availability routes are involved.

The Internet is also being heralded as the user's best option for access to increasingly large amounts of information. This is true, but at present the access mechanisms to the data are relatively new and the indexing and classification of the information is not done using any universal system. This presents a chaotic picture which tends to deter new users of the Internet and increase the cost of access with all but the simplest of queries. The problem is summed up in the cartoon in Figure 2.16 which is reproduced by kind permission of *The Guardian* and David Shenton.

2.5.3 Example applications

The typical applications that are used on the Internet are the four listed in section 2.5.2. Both electronic mail and file transfer are non-time-critical so present few problems in terms of the rate of data flow in the network. However, WWW pages use

'Stick That in Your CD-ROM'   by David Shenton

Figure 2.16 Information on the Internet

hypertext and the HTTP protocol and although this is notionally also non-time-critical the method of use as an interactive information access tool requires some degree of time-critical behaviour from the network. For example, a WWW page could be set up to display goods from a shop for on-line browsing using text, image and even sound and video if the goods needed to be heard or seen in motion. While the text presents no problem in terms of the amount of information it presents, images can be large data objects if the resolution or colour quality are high. Similarly sound and video increase, yet again, the data transfer requirements of the application. Therefore this type of application needs to be tailored to suit the network designated for access to it as use of a low speed network to access large data would increase the time required to browse the information and inevitably reduce customer interest.

2.6 THE INFORMATION SUPERHIGHWAY

There has been much talk about the information superhighway in the press and elsewhere. This has led to much speculation as to what an information superhighway actually is and when it will be available for users to communicate.

2.6.1 What is it?

The information superhighway was originally a term coined in the USA by vice-president Al Gore to described the National Information Infrastructure project. This aimed to build a high capacity backbone for a national network. Since this original concept the term has been used in other countries to refer to various national developments and speculations concerned with high speed data networks. Also the term has come to mean the next generation of the Internet when it becomes a high capacity network. So generically, the term now means any high capacity communications infrastructure.

This can take many forms and much of the information superhighway installation will go on unseen, but the widespread provision of cable TV networks that is happening in many countries will be a part of the infrastructure, albeit mainly for one-way entertainment provision rather than at this stage a true multimedia communication network.

2.6.2 When will it arrive?

As to when the information superhighway will arrive is a difficult question to answer. Firstly, elements of it are already in place, for example fibre and cable to industry, the home and education. Some, of it is about to be completed, such as high speed interconnections between local communications networks. However, other necessary features and conditions are not yet in place to make the whole system a reality. Commercially there is little profit outside entertainment so this is seen as a priority by service providers, and the provision of further services will depend, to a large extent on what is happening on the Internet. So there is a classic chicken and egg problem. Services for the high speed networks need to be demonstrated to be viable, but can only do so on low capacity current day networks. There are some pilot projects on high speed networks and these may prove useful, but providers need to see the commercial viability before any investment in services will take place.

So the information superhighway is starting to happen but it may be delayed while we all watch television!

FURTHER READING

This has been a very brief introduction to the areas of communication technology that will form the basis for further discussion later in the book. To gain more insight into this complex area the reader is directed at the author's previous book:

Andy Sloane, *Computer Communications: Principles and Business Applications*, McGraw-Hill, 1994, ISBN 0-07-707822-5.

If a broader treatment is required that includes detail of individual protocols then the following covers nearly all the subjects of interest.

Fred Halsall, *Data Communications, Computer Networks and Open Systems* (3rd Edition), Addison-Wesley, 1992, ISBN 0-201-56506-4.

For reference only, the following covers most of the relevant topics.

The National Computing Centre, *NCC Handbook of Data Communications* (2nd Edition), NCC Blackwell, 1995, ISBN 1-8554-002-9.

EXERCISES

1. If 16 users share a communication channel of 2 Mbps calculate the minimum channel capacity that each could use.

2. A communication consists of a video stream of 25 frames per second and an audio stream of a single mono voice channel.

    If each frame consists of 160 by 120 pixels of 256 colours and the audio is sampled at 8000 Hz with 8 bit sampling. Calculate the total capacity required to transmit the data in real time. Assume a 5% overhead for control information.

3. Calculate the capacity required if the communication in question two has a reduced frame rate of 15 fps and is reduced to monochrome in place of colour.

4. What compression ratio is required if this communication needs to take place on a single ISDN B channel?

3

MULTIMEDIA INFORMATION

Summary: The nature of information. Information types and structure, media types, properties and characteristics. Linkages between types. Relation to traditional methods of using media.

3.1 INFORMATION

Before the various media are described and discussed the nature of information itself needs to be outlined in more detail. To be able to communicate some information it must first exist or be created. This is often the result of some event taking place. For example, the result of a horse race is only a certain piece of information when the result is ratified by the officials at the racecourse, i.e. the events that have occurred are (1) the horses have raced and finished in a certain order (2) the race has been judged to be fair by the rules of the race and (3) the officials have agreed a result. At this point the piece of information that a particular horse has won the race is used for other purposes after it has been communicated in some way to any interested persons.

There are, therefore, two parts to this process. Firstly, the information must exist and, secondly, it must be communicated. These two are inextricably linked. Information is of limited application unless it is communicated or stored in some way to be communicated later. This ignores the scenario when a piece of information is originated by a process and is not used outside that process.

The method used for the communication of information will also determine the exact nature of the information transmitted. Although it may be possible to include all the details of a particular event in any form of communication it is not always possible to do so within other limits such as time or space requirements of a communication.

To use the example of the horse race again will illustrate this point. The events that need to be communicated cover a number of areas. Firstly before the race, many people including gamblers, racehorse owners, trainers, jockeys, bookmakers and racecourse officials will need to know some information about which horses are entered in the race or what the status of the race is, i.e. prize money etc. Although the information from this particular race may be summarised by the 1-2-3 of horses that finished, the total information package will include all the other details of the course, the jockeys, how each horse ran in the race, the finishing time, the distance between each finisher etc. To communicate all this information requires a number of types of information. Firstly, a video without sound would provide some information, each horse should be identifiable from the number it carries and the colours worn by its jockey identifying its owner. The time it takes to run the race should be able to be calculated, but it may not be possible to gather other pieces of information direct from the video. To give an indication of the state of the racecourse, the "going" would normally be described using a standard terminology, but this would be difficult to ascertain from just video. A voice or text communication would be best for this and other information related to the race. For the severely interested gambler, the previous races run by the various horses will have some interest and the actual state of each horse during the race is important to monitor for future events. All of this information may be carried using video, sound, text, still pictures or structured information from a database.

The various media types that are used to communicate information are only the available channels used for communication. They are not the information itself. Information is independent of the communication process until it is communicated. Then, each of the various media types put constraints upon the information generated by an event. So, if an event is to be communicated accurately a number of different media will need to be used or a compromise on the exact information that is communicated will need to be taken.

3.1.1 Types

Information exists in a number of forms. Chiefly among these are the distinctions already made between analogue and digital information. It is worth stating again that analogue information can be approximated very closely using digital techniques and the resulting digital information is often as usable as the analogue form. How much difference there is between the two forms will depend on the nature of the digitisation process and the original information. For the purposes of this book the analogue form of information will be largely ignored as the basis of multimedia and the communication of multimedia information is mainly based on digital techniques.

Two useful examples that illustrate digital-analogue difference are text and image.

Example 1

Text

Text consists of words written on paper. Each word uses symbols (an alphabet) to convey the information contained in the text. The original form of text was as handwriting on parchment or paper and then printed text in documents and books. All these are analogue forms of information. Each character can vary slightly from the standard form and the information is affected little. The digital form of the information is as used by computers in a format such as ASCII where each standard character is represented by a specific bit pattern. This loses some information from the written form but not usually a significant amount.

Example 2

Image

Images consist of a representation of visual information either in the form of a photograph or some other image form. The analogue form has changed from paintings to photographs over a number of centuries and the quality of two-dimensional images is now very high. These are all analogue information forms, with an infinitely variable colour balance and other parameters. The digital version requires that the image be sampled in some way and this will introduce differences between the original analogue version and the digital version. The sampling frequency (number of pixels per mm, in each direction) and the colour representation (number of bits per pixel) will each approximate the original by changing the shape of edges in the image and changing the colour of individual pixels. This effect will hardly be noticeable on a high resolution digitisation but can become very obvious if the display is only capable of low resolution with a few different colours.

The rest of this section will be mainly concerned with digital information, its structure and properties.

3.1.2 Structure

The fact that information has structure has been illustrated already in section 1.2 and above. When an event occurs there are a number of ways of capturing the information for communication or storage. These are related since the information contained in the event is associated with the event not the method of capture. So, consider an event, the tennis match again. The event is the final of a tennis tournament, such as the US Open or Wimbledon.

1. There is a certain amount of background information available about each player from previous events etc. and this adds to the total information for the match. (This is the context of the information.) An example is that the tennis match takes place in a certain tournament between two players who have beaten other players to reach the final round. Or that the rules of this tournament require that the winner is the first to win two sets.

2. The rules of the competition are as laid down by the organisers usually by taking those of the sport's governing body. These govern the information to the extent that they provide the structure to score and assess the event.

3. The event (the tennis match) can be broken down into sub-events (sets, games, points, rallies, strokes etc.). All events, except those that contain little information can be broken down in this way although the structure of tennis helps to a certain extent.

Now more precisely. The game can be represented in text by the following approximations.

	Set 1	Set 2	Set 3
Player A	6	2	9
Player B	3	6	7

This is the overall score and is enough to decide who is the winner and will probably be the most information that most people will store after the event. This could, however, be recorded with more detail such as

Set 1	Game 1	Game 2	Game 3	Game 4	Game 5	Game 6	Game 7	Game 8	Game 9	Score
Player A	Won	0	Won	Won	15	Won	Won	15	Won	6
Player B	15	Won	30	2x Deuce	Won	30	30	Won	3x Deuce	3
Server	A	B	A	B	A	B	A	B	A	

This now contains more of the information about the first set and the score at the end of each game can be seen. There is still much missing. However, from this information anyone interested in tennis would infer that player B did not play as badly as the may be inferred from the overall score. In fact the information appears to show that where player B lost a game he did so by a small margin whereas player A lost by much larger margins in the three he lost.

The next level of information would be much more detailed containing information about each point played and who won it. Then a further level could detail the nature of the shot played to win each point and so on. There are a number of levels of information below those illustrated that contain so much detail that they are very difficult to encode as text, and they take a large amount of room to do so especially for an event of this sort where new information is being generated with each hit and as each second goes by.

What is important to remember from this example is the structure that is inherent in all information. i.e. that there are many different levels that can be used and what is a difficult process, when using technological solutions to information transfer, is to use the appropriate level of information from its structure to communicate about an event. To a certain extent multimedia allows this problem to be reduced, firstly, because a number of different levels of information can be used and encoded in different appropriate formats and, secondly, because the user can choose which level is most appropriate to their needs.

When information is used and manipulated by computer systems there are often structures imposed on it by the system that is being used to store it. The diagram in Figure 3.1 is a generalised form of the structure of a computer and communication system. The various levels represent different views of the information stored in the system which can be extracted by processing the data stored in it or communicated to it.

The structure of the information contained in the example above is similar to the structure used to store or communicate information in computer systems. The high level information can be extracted from the lower levels of information which are eventually stored or communicated as bits or binary information. At this level there is no structure.

Figure 3.1 shows one analysis of a computer system. At the top level the user interacts with view of the information stored within the system that is determined by their own use of the system, this may be some structure imposed by a particular business use or home requirement, depending on the application. Then the information system view is the level at which the information is retrieved before being formatted into the user view above. The application view is dependent on the particular application being used, it could be a database or spreadsheet. Each will impose some structure on the data (although it may be by user choice at the design stage). Below that there will be some structure imposed by the language which was used for the application. (This may not be apparent until an information system reaches the limits of the design.) Below are restrictions imposed by the operating system, the actual system used and finally the encoding of individual items of information into bits.

The same is true of stored data as of communicated data. There needs to be a structure imposed on the data to enable good communication as there does to enable

```
                    ┌─────────────┐
                    │  User View  │
                    └──────┬──────┘
              ┌────────────┴────────────┐
              │ Information System View │
              └────────────┬────────────┘
                  ┌─────────┴─────────┐
                  │ Application View  │
                  └─────────┬─────────┘
            ┌───────────────┴───────────────┐
            │  Language Architecture View   │
            └───────────────┬───────────────┘
              ┌──────────────┴──────────────┐
              │   Operating System View     │
              └──────────────┬──────────────┘
                    ┌─────────┴─────────┐
                    │   Systems View    │
                    └─────────┬─────────┘
                      ┌───────┴───────┐
                      │ Data logical  │
                      └───────┬───────┘
                        ┌─────┴─────┐
                        │ Physical  │
                        └─────┬─────┘
                 ┌────────────┴────────────┐
          ┌──────┴──────┐          ┌────────┴────────┐
          │   Storage   │          │   Bitstream     │
          └─────────────┘          └─────────────────┘
```

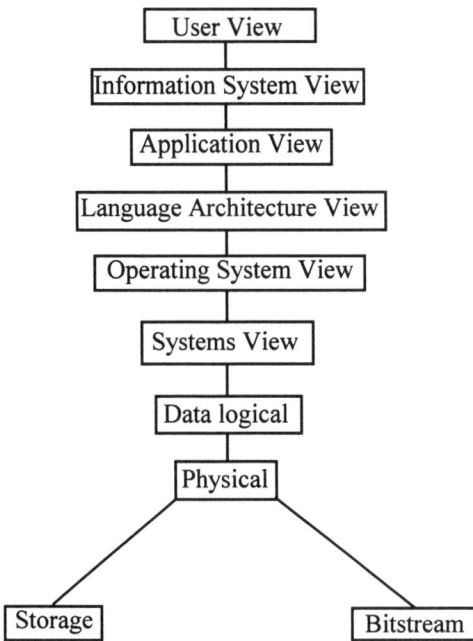

Figure 3.1 Structure of information used in a computer

retrieval after storage. Both require efficient encoding to enable the receipt or retrieval of the original information. As we have seen there could be problems with the information held in a computer system if the various levels are encoded inefficiently.

3.1.3 Properties

It has been shown that information can have a structure. Either from itself since it is created by an event and the associated context will generate structural information, or it can have structure imposed by systems that require information in specific formats.

    The first case is the more interesting. The second should be designed out of the system. Structure within information can have a number of effects, but most importantly the methods by which information is stored or communicated will depend on the structure it has. The most effective communication being for information that is appropriate to that structural level. The levels of structure in information are easier to encode in one way than in others. Although using the appropriate means of encoding may not always be the priority in a particular situation. If, however, choice is available, then an appropriate medium should be used. For example, the earlier example of the horse race allows video to be used for viewing the actual race, but the various statistics can be communicated as either text or structured information and a course map could be communicated as a graphic image.

3.1.4 Encoding

The structure of information, the available channels and the intended use of the information determines what the best system of encoding will be. The nature of the information to be stored or communicated should determine the encoding system used. So, if some information is generated by an event and it is to be communicated to a user of the information, then an appropriate decision should be taken to communicate the information in the correct form.

For example, if the result of a horse race determines the outcome of a bet, the result is the only information that is immediately of interest to the gambler. So any voice or text medium is adequate as long as the response of the system is within required time limits. Of course, video can be used and provides much more information but is not necessary for the strict purpose of transmitting a result. However, if the information is for the racehorse trainer or owner to use to judge performance, it may well be that video would be an appropriate medium to use.

Choosing the appropriate medium may depend on a number of factors. Many of these will be the result of a communications analysis carried out to determine the characteristics of the available channels. This will determine the type of object that can be transmitted and the scale of complexity of that object. For example, the capacity of any communications link will determine the resolution of real-time video, real-time sound quality, or the time it takes to transmit a certain image.

Figure 3.2 is a slightly altered version of Figure 1.1. Structured information such as that kept in a database system has been included. The diagram shows the relative requirements of the various encodings of information for storage or communication purposes. If the capacity is not available it will either not be possible to communicate or it will need to be done asynchronously. A partial solution can be gained by considering compression techniques. These alter the actual encoding, or rather add an extra layer of encoding, but can enable large data objects to be communicated when the raw version is too large. Its usefulness will, of course, depend on the extent of the compression possible, and the time it takes to compress and decompress the information. Section 3.2 will survey the area in more detail.

3.1.5 Multimedia

The difference between multimedia and the various single media encodings of information is mainly in the combination of media used for a particular communication. The capacity requirements of each medium now need to be assessed and added together, whereas the information in this format should be chosen because it fits the structural requirements that are relevant. So for multimedia to be the

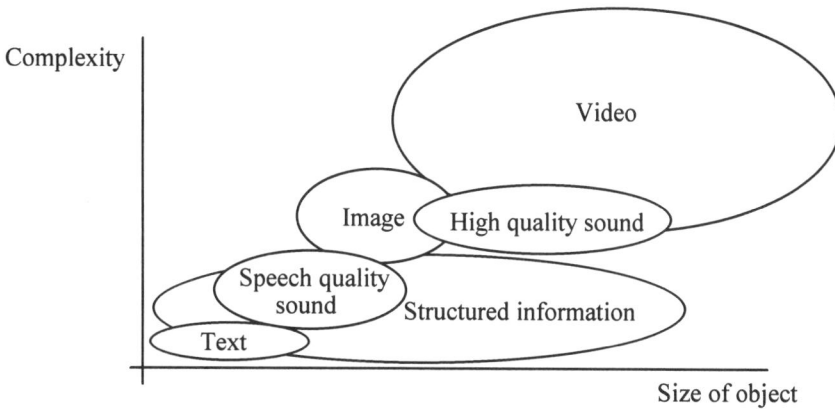

Figure 3.2 Encoded information objects

appropriate communication tool the capacity requirement must be met and the choice of media must justify the inclusion of each type of information. This may not always be the case as, in practice, the use of multimedia has been dictated more by fashion than information requirements. However, after the initial excitement, the analysis can show that using appropriate media for various information types can be productive. For example, a diagram such as a map can illustrate very effectively in less bits of data than the corresponding text. When the text of a journey includes a map it can reduce the capacity requirements, and make it considerably easier to understand.

3.2 MEDIA TYPES AND INFORMATION

The various media types that make up multimedia should all be familiar to computer users. The details may not be so familiar. This section is included to introduce the various media types and to give the necessary details for the later examples and analyses. What is not always appreciated is the difference between text and structured information, such as may be found in a database. While it is true that the encoding of free text and text fields in databases is similar, there are often other field types that are not encoded to the same degree of standardisation as text. Structured (database) information is, therefore, included in this section.

3.2.1 Text

Text is the simplest of digital information types to code and store. The nature of text is that it is composed of a limited symbol set (see section 1.3.1). This makes it simple to encode in a digital form. Each of the symbols can be encoded as a limited number of bits. Seven is used in ASCII. This is illustrated in Table 3.1

Table 3.1 ASCII character codes

Binary		000	001	010	011	100	101	110	111
								Bits	765
	Hex	0	1	2	3	4	5	6	7
0000	0	NUL	DLE	SP	0	@	P	`	p
0001	1	SOH	DC1	!	1	A	Q	a	q
0010	2	STX	DC2	"	2	B	R	b	r
0011	3	ETX	DC3	#	3	C	S	c	s
0100	4	EOT	DC4	$	4	D	T	d	t
0101	5	ENQ	NAK	%	5	E	U	e	u
0110	6	ACK	SYN	&	6	F	V	f	v
0111	7	BEL	ETB	'	7	G	W	g	w
1000	8	BS	CAN	(	8	H	X	h	x
1001	9	HT	EM	)	9	I	Y	i	y
1010	A	LF	SUB	*	:	J	Z	j	z
1011	B	VT	ESC	+	;	K	[	k	{
1100	C	FF	FS	,	<	L	\	l	|
1101	D	CR	GS	-	=	M	]	m	}
1110	E	SO	RS	.	>	N	^	n	~
1111	F	SI	US	/	?	O	_	o	DEL

Bits   4321

Each of the characters in the table can be represented by the seven bit quantity made from the three bits at the top and the four bits at the side in the order 7654321, so the binary version of the character **X** is 1000101. Even though only seven bits are used there are still some non-text characters in the ASCII set. These are mainly historical and are connected with communication.

So it is relatively easy to represent text with binary digits, but the table only applies to English text and any language that uses the same symbol set. There are, however, variations for international use that have been standardised. ASCII is also known as International Alphabet No. 5. There are other languages that use completely different character sets and Chinese and Japanese languages are good examples. These use many more symbols.

To solve these problems, there are now moves to standardise on a universal code for characters of all types using a 16-bit format known as Unicode. This will allow 32768 different symbols, which should be enough!

To illustrate the size of a text document the following example is indicative.

Example

A sheet of A4 paper can contain about 60 lines of type with 96 characters per line using a small type face. This is

$$60 \times 96 = 5760 \text{ characters or } 40\,320 \text{ bits.}$$

If this is communicated over a channel with 9600 bps capacity it would take 4.2 seconds to transmit (assuming no added error control information).

Part of this efficiency is due to the structure of textual information. There is a large element of redundant information in text which is used in the subsequent ASCII encoding. This allows a relatively simple encoding of the symbols to return the original information.

3.2.2 Structured information

Information that is held on databases is stored with a pre-determined structure that has restricted the variety of the data. This has implications for the communication of the information stored and the use that can be made of it. There is also the problem of physical storage schemes being non-standard and leading to misinterpretation of data in communication unless appropriate standardisation is used. For example the first of these problems is apparent in a situation where one application (on computer **A**) communicates with another (on computer **B**). If **A** stores the data as 16 ASCII characters and **B** stores the same data as 24 ASCII characters, **B** cannot increase the data to fill the field and make use of the extra characters, likewise **A** will not be able to handle data from **B** and retain all the relevant information contained in it. The second problem may occur when numerical data is stored in a different format in the two systems. All these problems are solvable but the user or system developer needs to be aware of differences in data storage if errors are to be avoided.

Structured information can be small or very large depending on the application used and the system used to access it. Two examples can illustrate this, the first is of a similar scale to the text example above.

Example 1

A database has the following table structure

Field 1	Field 2	Field 3
Record identifier	Name	Telephone number
Numeric 5	Character 12	Character 13

If a list of entries is communicated, there will be approximately 60 per page. So a single printed page version of some typical entries may look like the following.

Identifier	Name	Telephone number
00012	A Smith	0121 345 6789
00034	J Brown	01884 566779
00045	A Thomas	01335 342343

A typical transfer of information of 60 records from a table would need 50 transfers of 15 characters plus a numeric which could be a 16-bit quantity. Which gives

$$60 \times ((15 \times 7) + 16) = 7260 \text{ bits}$$

or a transmission time of **0.75** seconds using a 9600 bps channel. (Again, assuming no added error control information.)

Example 2

A larger example can be calculated from a typical national database. A country has 5 million drivers on a database, with each driver's record containing at least the following basic information:

Driver number	Driver name	Address	Entitlement classes
Character 12	Character 20	Character 100	Character 10

(Please Note - This is a fictitious example used for illustration purposes only)

Transfer of all these basic records from the database to another would require a transfer of 142 characters per record or

$$5\,000\,000 \times 142 \times 7 = 4\,970\,000\,000 \text{ bits}$$

which would take **517 708** seconds on the same 9600 bps line. That is **5 days, 23** hours, **48** minutes and **28** seconds! Assuming no errors. This is clearly a candidate for a faster speed communications link.

At **2 Mbps** the data would take **2485** seconds or **41** minutes **25** seconds.

It is clear that the communication of structured database information can cover very small to very large transfers of data, but typically the information is relatively simple. The users need to be aware of the amount of data being transferred and the consequences of particular queries to remote databases. For example, in the above national driver database if a user asked for all the data on drivers whose name is Johanson there may be a few thousand entries, but if they made a mistake in the query and missed out the name the system may try to send all the records in the database. As can be seen this may take some time. So incorporation of database queries into multimedia communication has two sides: (1) it is a succinct expression of simple information that can be queried for matching entries (2) it can produce massive amounts of data if errors are made in compiling queries.

3.2.3   Audio

Audio is more complex than the media discussed so far. There are a number of standards which are common among audio systems and applications and machinery often needs to encompass more than one standard. There is also a range of applications that use sound of different quality levels, and cannot necessarily cope with higher quality audio because of other factors, such as capacity limitations. There are two commonly used standards that are worth further evaluation. High-quality stereo audio and speech-quality mono audio. These have very different communications needs when being transmitted as digital information.

Example 1

High-quality audio (Cd-standard)

This uses a sampling rate of 44.1 kHz with 2 channels of 16-bit samples.

In real-time use this requires a capacity of

$$44\ 100 \times 16 \times 2 = 1\ 411\ 200 \text{ bps}$$

Or if a one second burst of audio is sent along a 9600 bps link it will take 147 seconds to transfer or 147 times real time.

Example 2

Speech-quality audio

This uses a sampling rate of 8 kHz with 1 channel of 8-bit samples.

In real-time use this requires a capacity of

$$8000 \times 8 \times 1 = 64\ 000 \text{ bps}$$

Again if a one second burst is sent down a slow 9600 bps channel the communication would take 6.7 seconds or 6.7 times real-time.

The difference that marks audio from the previous two media types is the real-time nature of its normal use. Sound is a common medium for human communication and it is common to hear analogue systems deliver voice and other sound communications in real time, e.g. on the radio or TV. This ability to deliver audio in real-time is a common requirement of multimedia systems.

In multimedia communications systems this then requires enough capacity to allow this but at the same time as not reducing the transfer of other elements of the media mix to an unacceptably slow level. It is, therefore, crucial to use the correct quality level of audio in an application that requires real-time communication and not to incorporate the best quality of audio regardless of use. Calculations will need to be carried out concerning the relative proportions of different media used in a communication or the limits on their use, if they are exchanged along relatively slow communication channels.

3.2.4   Image

The image medium is more diverse than others since it can include both photographic-type images and the graphic images of design-type information. These can both be represented in digital form as bitmapped information, but the graphic image can also be represented in a metafile format that is a standard form of information used to denote how the original graphic image can be reproduced from a set of "instructions". At present the main concern of this section is the first type of image — the bitmapped photographic image. If the user needs to manipulate the information contained in a graphic image they will need some form of information that can be manipulated. It is not a simple problem to recover and manipulate information from a bitmapped image. An example of each of these forms of information is shown in Figures 3.3 and 3.4.

Figure 3.3 A bitmapped image

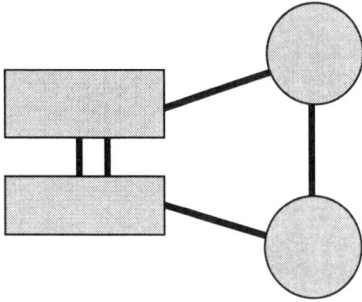

Figure 3.4 A graphic image

Example 1

Bitmapped images

In Figure 3.3 the bitmapped image is a scanned photograph. The whole photograph is sampled into pixels at a resolution of 160 x 120 pixels. Each pixel is then represented by an 8-bit colour number. So this image is

$$160 \times 120 \times 8 = 153\ 600 \text{ bits}$$

Using the standard speed communications line of the previous examples (9600 bps) this image would take **16** seconds to transfer.

It is, however, a very small version of the image and a full screen version would be 800 x 600 pixels (or more) which would take 400 seconds to transfer, that is a full 6 minutes 40 seconds.

Example 2

A graphic image

The graphic image in Figure 3.4 is composed of a small number of elements that were drawn using a drawing package. These are stored in a different format to the image in Figure 3.3. Each element is stored as a set of parameters and co-ordinates. This allows a much lower requirement of information to describe the object. In this case there are nine elements, that is, two rectangles, two circles and five lines. Much will depend on the system used for the encoding of the graphic elements but a simple system could use about 64 bytes to represent each element and this would give

$$9 \times 64 \times 8 = 4608 \text{ bits}$$

On the example 9600 line this would take **0.48** seconds to transmit. In reality this would probably be an underestimate since there is usually more information included with the image than just these bare bones, but even if the figure is doubled it is still a much simpler format than the straight bitmap.

There are many different forms of image coding that will be discussed in the next chapter. At present the above analysis is sufficient.

3.2.5 Video

Video is a form of moving pictures that relies on the human inability to distinguish the difference between images changed at a fast rate and real movement. The system was first used in cine pictures where the rapidly changing image gives the impression of movement, but it is really only a sequential set of images displayed at a fast rate. The standard rate is between 25 and 30 frames or images per second. This gives the impression of smooth movement between successive images and the human eye is unable to detect the individual images. As the frame rate reduces the images become more distinct and the viewer can detect the movement becoming jerkier and less fluid between frames. Under about 15 fps (frames per second) the image seems a little unreal but still usable.

The data rate needed to transmit video will, therefore, depend on a number of factors. These are

1. The frame rate
2. The size of the display
3. The resolution used
4. The number of colours in the image

In practice the resolution and size of the display are used in conjunction by giving the image size in pixels. This means that different viewers will have different sized display windows depending on the resolution they happen to be using on the display screen. For the following example calculations this scheme shall be used as it simplifies the processing.

Example 1

Low-quality video in a screen window
Consider a system with the following parameters

15 fps refresh rate in a
160 x 120 pixel window with
16 colours or grey scales (4-bits per pixel).

This would give a usable image in ordinary circumstances but would not be adequate for high-definition use. Total data capacity of the image only (sound is separate) would be

$$15 \times 160 \times 120 \times 4 = 1\ 152\ 000 \text{ bps}$$

This is about 1 Mbps for a very small image at low refresh rates. Obviously compression will reduce this figure but it does show the amount of data that is needed to provide video over a communications link. One of the useful features of video is its ability to mask errors. If data is corrupted in transmission, the display may have a few minor problems displaying the image but, since there is a new frame arriving every 1/15 s the effect will be minimal unless a massive corruption of data occurs. Each frame consists of 76 800 bits and burst errors of 10 000 bits would only cause slight disruption to about 1/8 th of a frame.

If the 9600 channel is used to transfer the video data in this example it would take **120** seconds to transfer just one second of video and so would only be useful for short video clips that can be played off-line later. Even with compression there would be considerable difficulty in reducing the data enough to make the use of a 9600 channel worthwhile.

Example 2

High-quality full-screen video
A system with this following parameters could be considered as a high-quality video format:

25 fps refresh rate for a
800 x 600 pixel screen with
65 536 colours or grey scales (16-bits per pixel).

This gives

$$25 \times 800 \times 600 \times 16 = 192\ 000\ 000 \text{ bps}$$

or 192 Mbps. Clearly the same applies as in the case of the low-quality video, i.e. compression will reduce this considerably without loss, or even more if lossy compression is used. However, for good quality video a high bit rate is inevitable.

It would be patently ridiculous to try to send any of this quality of video over a slow speed link such as the 9600 bps used in the other examples. In this case it would take 5.5 hours to send one second of video or 20 000 times real time!

Video is the most data rich medium in common use today, mostly this stems from a high redundancy factor. There is little change from one frame of video to the next and if each requires sending the utilisation of the channel is not efficient. With compression the redundancy can be reduced by various techniques used to filter out unchanging (i.e. redundant) information and sending the differences in frames. These techniques are common and will be described in Chapter 4.

3.3 MULTIPLE MEDIA

The use of the term multimedia has been come to be used when the various media type described in the various sections of 3.2 are used in conjunction with one another. There has always been some degree of use of various media in this way and some of these uses have been relatively common even before the use of computers allowed the digital versions to become established as multimedia. The change between what has been possible in the past and the current multimedia systems is most noticeable in the ease with which these different types can now be manipulated to form multimedia documents etc.

3.3.1 Typical use of multiple media

Although the use of sound with video could "technically" be described as multimedia it is unusual to find it so described. This is probably due to there always having been a sound portion to video communication since the beginning of television. This combination will be excluded from the multimedia category.

There are, however, a number of combinations of media that have, in the past, been used in various physical manifestations. These will be described and discussed in the following set of examples.

Example 1

The slide show (sound + image)
A traditional slide show consisting of projected images and a voice sound track has been used as a means of describing events and journeys for a number of years. The advantages of using these media are simply that it requires little equipment to produce the slides and the talk and is easily portable to different venues to display the end result. It also allows insertion of images to illustrate extra points, although a measure of prior planning needs to be done if a complete record of events is to be recorded. It also lacks movement which may be

essential to certain events, e.g. motor races or similar events. It does, however, have a relatively widespread means of production in that slides can be produced easily and cheaply. Slides can also be combined together to form programmes other than an original event.

A slide show is also common in digital form with images being manipulated into forms to suit the presentation and the audio soundtrack may be synchronised with the image display with a single clock since there is no time sensitive element to a simple image.

Example 2

The picture album (image + text)

A traditional picture album will be familiar. A combination of images and text describing or annotating the pictures is a common experience. Picture albums are used to describe events or sequences of events or simply to record or store images of interest.

The digital version is effectively a database (or even a sequential file) of images and related text. This is simpler than the previous example since there is no synchronisation needed and many packages from simple word processors to sophisticated database managers can manipulate text and image to display information in this form.

Example 3

The illustrated newspaper or book (text + image)

The main difference between the picture album and the illustrated book or newspaper is that the main emphasis is on text in the latter case, rather than on the images as in the picture album.

In the digital version of an illustrated book the text will normally be in free form with embedded images to illustrate the text. This is common enough in everyday life to be familiar to all readers. This is the model for this book!

Example 4

The picture library (image + structured information)

On a larger scale than example 2 is the picture library, where many thousands of images are stored with cross-references to events, people, places and other pertinent categories. In a traditional setting these would be printed images with index cards to link together material from different sources that could be used in a single setting. As a possible application of the picture library, a journalist might want to illustrate a story with pictures of the people or places involved and include some background pictures of either the people at some other time or a previous view of a place. As an example of this assume a person becomes a member of parliament, a journalist might want background information from that person's past, pictures of the election campaign, pictures relating to relevant issues in the campaign or a standard picture of parliament itself. There are many possibilities.

Digitally this can all be done via a multimedia (or, specifically, picture) database where the images are classified according to content and context into categories with cross-referencing and indexing being heavily used to obtain as wide a range of images as possible.

Example 5

The language course (sound + text)

Learning a language has long been accompanied by sound. It is difficult to learn pronunciation without hearing the words first, although possible it is not an easy way to learn. The most common combination until recently has been the text and audio cassette. This enables a language to be studied via the text and linked with the pronunciation via the audio. These packages are very common and have formed the basis of simple language learning for many years.

With the digitisation of sound and text, the language learning package can be updated to be more interactive. The digital version allows users to find pronunciation for passages and words in text by having cross-referenced databases of sounds and text. This can also be expanded to aid learning by introducing image or video.

Example 6

The self study package (video + sound + image + text)

There have been many types of learning package produced over recent years. Many of these have been very well designed and are highly structured approaches to learning or training. A lot of careful planning needs to be done to ensure that difficult concepts are described and taught in a way that is relevant to the subject and the learner. These packages are, therefore, difficult to put together having books, videos and audio cassettes to mix and co-ordinate together.

Digitally this is simpler as long as all the material is available on a standard platform. With a relevant authoring package the various strands of the package can be linked together. In fact, there is a method of doing some of this via a communications network which will be described in Chapter 8, i.e. hypertext mark-up language, HTML. Although this is not exactly the same as the case envisaged in this example.

These examples have shown some of the diversity in multiple use of media and the traditional uses that have been made from multiple sources of information. Many of these have been in use for a long time, some are more recent. One factor is, however, worth noting. The availability of information in a particular form will normally lead to its inclusion in some multiple media package, so if the information is available it will be used.

3.3.2   Linkage between media

The examples in section 3.3.1 have shown how multiple media have been used together in various ways to form integrated packages for specific purposes. Much of what is done has derived because it was the only way to be able to include more diversity of material in different formats, but in some cases there are specific needs for lower diversity. For example, it is possible to produce a multiple media newspaper with videos and audio cassettes supplementing the text and image of a standard offering. It would, however, be very expensive to produce and the increase in cost to the consumer would not generate sufficient sales to warrant this type of extra inclusion, using traditional media techniques. There have been a number of different media used for more specialised applications similar to this. The publication of a music magazine now often includes an audio cassette or CD as an extra, although this is not always related to the content directly. Sometimes video cassettes are used in a similar way and advertising can now be found in all these different forms.

The linkage between different media has been outlined in the examples and these links will still be used in digital versions of multiple media. They should be easier to implement. Firstly, the all-digital version of the information will allow a single digital storage or communication device to be used and the problem resolves into how individual items are linked and synchronised.

Links between items can be of two forms. One where the link is part of the document and the item is automatically used. For example, in a newspaper where an image is include in the text. The second is where an explicit link is contained in the information stream. An example of this would be a link to a musical soundtrack in a text referring to a composer's work, i.e. play track 6 for example. Links of both forms can be found in many examples and are often used in real situations. Synchronisation is covered in the next section.

3.3.3   Synchronisation (e.g. audio-video, image-audio)

Mostly synchronisation is not particularly problematic, except in the area of video/audio synchronisation. Otherwise a simple technique based on deadlines can be used. This allows an element, such as an audio clip to finish by a particular time so that the rest of the programme can be progressed. If the audio is associated with an image display then the synchronisation only requires that an image change takes place at a particular point in the audio track. This is illustrated in Figure 3.5

Image 1   Image 2   Image 3   Image 4

Audio samples (x1000)   Time

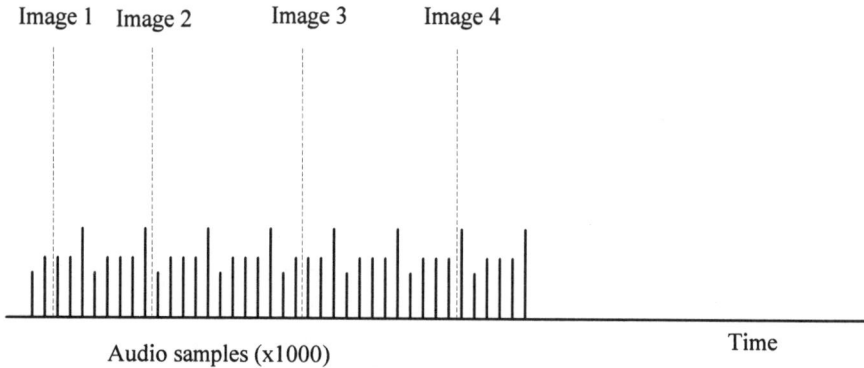

Figure 3.5 Audio-image synchronisation

The diagram shows how synch points can be used to time the display of images to particular points in the soundtrack. This type of synchronisation would be used in a slide-show type activity. Mostly an audio soundtrack would follow the image display and any synchronisation would be relatively simple.

In video-audio synchronisation there is a more difficult problem since the use of real-time linkage needs careful timing. The problem is shown in Figure 3.6. Here the video frames are shown with the associated audio stream. At a frame rate of 25 fps and an audio sampling rate of 8 kHz there are 320 audio samples to process per video frame. Possible synchronisation is shown at the beginning of each frame. However, there is usually no need to synchronise this often as the tolerance of the user to unsynchronised video-audio is higher than this and a rate of synchronisation of once every five frames could be used.

In practice, the problem is solved by capture and storage hardware and software in many cases, since an interleaving of audio and video tracks is used. In communications systems, this is not the case and techniques for synchronisation need to be adapted to the channel available. This requires the frame rate and sample rate to be adjusted to reach a level where both can co-exist on the channel and data can be sent at appropriate rates. For example, if the frame rate is increased the audio sample rate must be decreased in a fixed capacity channel. If the channel is variable, as in contention based networks, the video frame and audio sample rates must be both reduced or increased in line with available capacity.

3.4 USE OF MULTIMEDIA

Information in different forms has been used in many ways since communication was invented. The printing press saw the use of text spread rapidly, illustration was used from an early date to add information to the text and to expand on the information

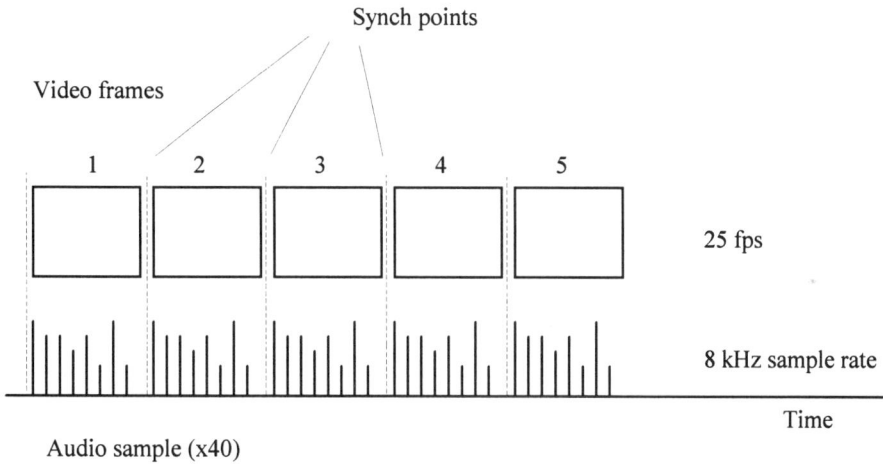

Figure 3.6 Audio-video synchronisation

contained within it. The invention of photography saw the widespread use of images and the advent of radio and assisted in the development of sound processing. Finally, moving pictures started with cinema and progressed to video. Recently digital information technology has allowed all these forms to be combined in single programmes for many purposes. This raises a number of questions which this book will try to answer.

3.4.1 Issues

The issues connected with multimedia vary with the application, but there are some core issues that are relevant to all forms of media.

Firstly, the problem of access to the media concerned. How is access achieved and regulated? Are there particular attributes that allow any interested user to gain access to the information or is it restricted in some way? The commonest restriction being that a potential user may not be able to buy the information and is at a disadvantage by virtue of economic constraints. This may be intentionally part of the access package devised by the owner of the specific programme involved. However, if the information is otherwise freely available but access is restricted by virtue of particular equipment then the question of appropriate use of the media for this purpose should be raised. For example, if a company wished to sell financial advice on share dealing via a multimedia network, the economic considerations are unlikely to affect the range of potential users. If, however, a government wished to use multimedia in a Health Education campaign it could be argued that only a small proportion of people who are likely to benefit from the information would be able to access it. The appropriate use of media is influenced by many factors as will be seen throughout this book.

A second issue which covers many media types is the control of the information provided through the medium. It may be that the content is controlled in some way by regulatory bodies with powers to control the information content of particular media, or there may be less restriction of content which then allows individual control to users, but at the same time does not restrict access to unwanted or unsuitable material. An example of this aspect would be the regulation of films in some countries where a censorship body controls what may be shown on film and video. The use of the Internet to access international services allows local laws to be circumvented by using overseas servers. The Internet cannot be said to have any controlling power determining content of any communication. What is clear is that the use of multimedia information is set to grow and the use of international networks does not allow any single nation control over the content or information. There are also many other factors affecting individual media and programmes of information.

3.4.2 Interfaces

The universal interface to digital information is the computer. This book is designed to be as neutral as possible in the type of computer discussed, but there are always individual differences that become apparent to users that make particular machines useful to their particular tasks whereas others may not be as effective. However, the current generation of computers and workstations have all similar characteristics that allow some degree of generality to be introduced to the discussion.

Firstly, to qualify for inclusion in any discussion on multimedia a computer must have some capability to display images at a resolution that allows adequate definition for the purposes of the application. A screen with at least 800×600 pixels and 255 or 65536 colours is now a minimum requirement. Secondly, an audio interface capable of processing 8 and 16 bit sound samples at rates up to 44.1 kHz sampling rate is now a norm for most multimedia machines. Finally, a multimedia machine must be able to process video in real-time at an adequate resolution and with a usable frame rate.

Without any of these qualities a computer could not really be said to have an effective multimedia interface. It will, however, be noted that when multimedia communication is considered, there may be aspects that are not so important since the usual determining factor is the speed of the communication link not the machine attached to it. If the information is to be processed after transmission, however, then the above criteria will apply.

3.4.3 User needs

A further criterion that needs consideration when multimedia use is proposed is the actual needs of the users for whom the system is intended. If the user's requirement is to exchange text information between colleagues on a number of sites of a commercial company, the use of email may be sufficient. However, in any analysis the possible options should be included in the initial analysis so that informed decisions can be made. If a network is installed that can handle the email and the provision of video links between sites is then required it is possible that the whole network may need to be upgraded, but with sufficient planning in advance an option to upgrade to higher capacity links could have been provided from the outset.

The planning of communication networks is a complex business, made more difficult by the use of multimedia since both overall capacity and guaranteed capacity are important to the user when time-sensitive information is being transferred.

3.4.4   Actual use and applications

Finally, the actual use of multimedia communication systems should be considered. The capability to achieve a particular level of service can often have a considerable effect on the spread of service provided. For example, if a network were to be installed on the basis of needing video links between users, this would require the capacity of the network to be able to cope with the projected average use of this facility. If the projection is wildly inaccurate then the extra cost of higher capacity may have an adverse affect on the other facilities provided on the network. This may cut down the individual service options allowed to users or reduce the number of other facilities available.

In terms of applications, users should be aware of the capabilities of the networks and communications links that are available. A lot of time or money can be wasted by using inappropriate communication links, and if the level of service required is not available to the user population a network may be regarded as a useless object when with a carefully planned installation it could provide a tool to transform the information access of its users.

FURTHER READING

There are a number of books now available on multimedia, one that will be of interest to the more technically minded is:

John F. Koegel Buford (Ed.), *Multimedia Systems*, Addison-Wesley, 1994, ISBN 0-201-53258-1.

EXERCISES

1. Calculate the maximum number of simultaneous users who can video conference on the following networks.

    a) a 16 Mbps token ring network
    b) a 10 Mbps CSMA/CD network

    Assume that the video conferencing is of the following specification

    i) 160 x 120 pixels at 15 frames/s with 8 kHz mono audio
    ii) 320 x 240 pixels at 20 frames/s with 8 kHz stereo audio

    State clearly any further assumptions made.

2. A company has an installed 10 Mbps CSMA/CD network with 50 users.

    Devise a policy, or code of practice, for use of video conferencing over the network so that normal work is not disrupted.

3. List the factors that might affect the actual use of multimedia in the following applications

    a) A video phone call
    b) A music store
    c) A scientific academic conference
    d) A medical consultation

SYSTEMS, TOOLS, APPLICATIONS AND STANDARDS

Summary: A brief overview of multimedia developments. A more general discussion of the multimedia system and the various tools and aids available for the manipulation of information.

4.1 MULTIMEDIA SYSTEMS AND STANDARDS

Since the beginning of the computer age there have been problems which stem from the variety of equipment and software that is available to the user. The biggest problem associated with the diversity of computing platforms and their use is the lack of universal standards applicable to all systems and users. There are essentially two viewpoints to this problem:

1. All diversity in computing is to be encouraged and any differences between systems can be accommodated by translation software or special devices.
2. All diversity in computing should be discouraged and standards should be rigorously defined and used.

While both of these positions are often firmly held beliefs by opposing protagonists, the reality is that a compromise between the two extremes is the way that the computing industry appears to be heading. However, what is often difficult to predict in the fast-moving world of personal computers is the likely uptake of particular standards and the possible inclusion of new and different hardware and software in the specification of systems for multimedia and communication.

So, while this chapter can take a snapshot of the current (1996) standards and systems, it will inevitably become dated in a short time. The power and facilities of typical personal computers are being improved every year and the difference between machines bought only one year apart can be significant, especially when the use of complex multimedia packages is contemplated.

Multimedia software has, to a large extent, been a focus for the continuing development of personal computers. There are now an increasing number of machines sold with CD-ROM as standard and the ability to use audio and video are now commonplace. Together, these developments have come to be known as a multimedia PC. There are detailed standard specifications for these and they are outlined in section 4.2. The use of multimedia is taken as a baseline in many of the applications being sold for entertainment purposes and the specification of machines to use the software increases with each new release. There is, therefore, an upward spiral of software and hardware specifications that are required by users to use the latest applications and enter the multimedia arena.

In communications the picture is less frantic, since many of the developments have taken a more leisurely pace into public consciousness. This does not mean that developments are not happening, but that the interested parties are not promoting their use as actively as the PC manufacturers are doing with the multimedia PC. Some of the developments have already been discussed, such as ISDN, ATM, video conferencing standards and so on. What is happening in communications is dependent on which country is concerned. Some are more advanced than others in their use of high-speed data networks, and the installation of comprehensive fibre optic networks linking users together.

The one aspect of communications that is significant is the near-universal agreement on standards that has been the hallmark of the industry for many years. This allows more effort to be put into specific solutions that will interwork with others from different suppliers. This situation is very different from the standards associated with computing where competing standards often appear at the same time and the resulting outcome is a large number of users with a "standard" piece of equipment or software that will only work with other identical configurations. Many users are forced to wait to see the outcome of these marketing battles before deciding which of the competing standards to actually use. What may be the inferior standard can often win out in this type of conflict.

It is, therefore, interesting to note that in the multimedia area there are a number of standards that have been developed by interested groups to provide a basis for future applications in this area. Some of these are particularly relevant to the communication of multimedia information and will be outlined in detail in section 4.3.

What is clearly, theoretically possible is the ability of any user using any computer to communicate some multimedia information to another user using a different type of computer. What is practically achievable is limited, at present, but with increased emphasis on the standardisation process the theoretical may become the commonplace in a short time.

The following discussion concentrates on the PC as the platform used most widely in the world. There are, however, sizeable groups of users of other hardware, such as Apple Macintosh™ and various Unix™-based workstations. While these are significant in the overall development of multimedia the user-base is much smaller than the standard PC and this is therefore the platform considered most within this chapter.

4.2 PERSONAL COMPUTER STANDARDS

The widespread use of the IBM™-compatible PC has made standardisation a much easier task in the personal computer area. While there is competition from other formats the market share of the compatible PC is high and this allows a degree of standardisation for users that results in lower prices and wider choice of devices and software. However, at the beginning of the multimedia revolution a number of interested parties formed the multimedia PC working group under the auspices of the Software Publishers Association and published detailed specifications for multimedia PCs.

The MPC specifications are based around the CD-ROM as the multimedia carrier rather than around a communications-based idea of multimedia, but they do form the basis of most PCs that are used for multimedia communication, and they provide a standard which is being developed as the hardware and software changes adding increased capability and function.

It is, therefore, likely that the MPC standards will continue to be important as long as the IBM™-compatible PC is as dominant in the marketplace as it is currently. It is also interesting to note that the MPC standards have largely defined multimedia for the PC, which could easily have become a diverse set of incompatible pieces of hardware and isolated software tools had the effort not been made to fix the specification. What is further noticeable, is that the MPC standards encapsulate other standards that have become industry-wide standards through use, the *de facto* personal computing standards.

As will be seen in later chapters the use of standards is of crucial importance to individual applications, especially in communications where the ability to communicate is dependent on the adoption of some standards by the communicators. This has already been outlined in Chapter 1 but is particularly important here. Standards in the communication process can be used in a number of ways. The definition of communication protocols can allow any manufacturer to adapt or develop hardware and software to interface with the communication standard and provide users with a communication tool that will promote widespread use. If standards are disregarded either the device will disappear into specialised markets where compatibility is unimportant or, as can happen, become the *de facto* standard amongst users. The latter course is not always an option for smaller manufacturers. To reduce the uncertainty of non-standard equipment in the multimedia area the MPC specifications were devised.

4.2.1 MPC standards

The original MPC specification was first published in 1991, this was updated as MPC2 in 1993 and in 1995 MPC3 became the highest level of specification for the PC. The original MPC is now too dated to be of much interest, and MPC2 has been largely superseded by the increase in processor speed and disc capacity that has occurred since 1993. The 1995 MPC3 essentially encapsulates the state of multimedia at that time. A brief synopsis of the important aspects of the MPC2 and MPC3 specifications are listed in Table 4.1

While most of the specification is concerned with the ability to run multimedia software from the CD-ROM, the standard MPC2 or 3 compatible PC will allow a user a degree of certainty about compatibility and the capability of the system. In turn, this will allow the output from a multimedia communication channel to have a standard of display on such a device that is uniform across similarly standardised devices.

To calculate the level of communication channel supported by this specification is relatively straightforward. Any communications input into the system would need to be capable of feeding the video and audio capabilities at least and this provides a basis for the calculation. A small text input has been added also but this makes little difference. The following calculation assumes that no compression is used.

Video	352×240 at 30 fps $\times$ 15 bits/pixel	=	38 016 000 bps
Audio	8 kHz samples of 16 bits (stereo)	=	256 000 bps
Text	up to 9600 bps	=	9600 bps
	Total	**=**	**38.3 Mbps**

Clearly this is considerably more than most of the currently used technology is capable of supporting for individual users and so a machine capable of this speed and resolution should be more than adequate for the multimedia communication tasks over relatively slow networks. However, with a typical compression ratio obtainable from compressed video being in the order of 30:1 the bit rate reduces to a manageable 1.2 + Mbps. (Further details of this compression are in section 4.3.4.) So for an MPC3 system to take full advantage of its own capabilities on a network using multimedia communications a communication input of 1.2 to 3.0 Mbps would be required. This is more than most current LANs would allow, consequently new communications technology is seen as the way forward for multimedia communications.

Table 4.1 MPC 2 and MPC 3 specification

	MPC2	MPC3
Processor	486SX (25 MHz)	Pentium (75 MHz)
Memory	4 Mb RAM	8 Mb RAM
Storage	160 Mb hard disc drive	540 Mb hard disc drive
Audio	16-bit digital sound	16-bit digital sound; wave table
Graphics	1.2 Megapixels/s delivery given 40% of CPU bandwidth	Colour space conversion and scaling capability; Direct access to frame buffer for video-enabled graphics sub-system with a resolution of 352×240 at 30 fps (or 352×288 at 25 fps) at 15 bits/pixel, unscaled, without cropping
Video	Not applicable	MPEG1 (Hardware or software) with OM-1 compliance; all CODECs to support a synchronised audio/video stream with the resolution above without dropping a frame
CD-ROM drive	300 kbps transfer rate 400 ms max. average seek time	600 kbps transfer rate 250 ms max. average seek time
User input	101-key IBM-style keyboard Two-button mouse	101-key IBM-style keyboard Two-button mouse
I/O	MIDI; joystick; serial; parallel	MIDI; joystick; serial; parallel
System software	Windows 3.0, plus multimedia extensions, or binary compatible	Windows 3.11, DOS 6.0, or binary compatible

Much of the work carried out in multimedia compression has been directed at producing data compression techniques that can be used on CD-ROM as this medium is seen as the starting point for multimedia presentations and multimedia information. There are also compression techniques that have been developed specifically for communication, there are differences but the need to compress the data enough to use the standard data rate of CD-ROM has been useful in the communications area.

4.3 MEDIA STANDARDS AND COMPRESSION

A standard PC specification is only part of the whole picture. The other part is the standards for individual data types or media within multimedia so that devices can process the information in a communication. The following sub-sections look at each of the data types in turn and describe the various common standards used for each, including the compression standards used in the types where raw bitmaps produce large data objects, i.e. image and video.

4.3.1 Text exchange

There are two opposing modes of using text in a communication, either as the sole encoding used when a data transfer can be done at speed for later use i.e. off-line reading, or as a lower volume communication which can be read as soon as it is received. There are also points between these two extremes. Figure 4.1 shows how the usable data rate is affected by the mode of reading for a text document.

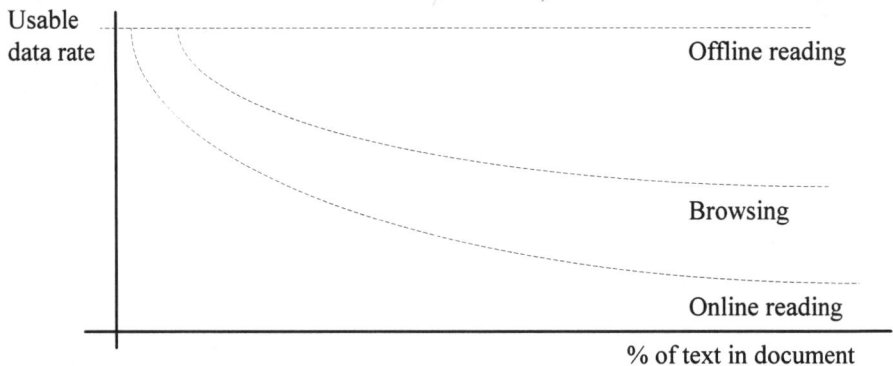

Figure 4.1 Text use in communicated documents

In simple off-line reading the usable data rate is unaffected since all the text will be stored for later use. This is convenient when data is known to be useful but is too large to read on-line. The other extreme of on-line reading is useful for small documents or large multimedia objects with small amounts of text which require immediate reading. In-between the two extremes a browsing mode is shown where only partial information is read on-line and the rest left until a complete download is done. This is a common way to check that communicated documents are correct when being received and also to find documents on a network. This diagram illustrates the use of text in communications. When a document is largely text the modes will differ considerably, but with reduced amounts of text the differences reduce.

In numerical terms text data only needs a few hundred bits per second to provide enough text to occupy an on-line reader fully. As can be seen from the diagram the usable data rate in this case is the lowest of all the cases.

Text has already been discussed, in part, in section 3.2.1 where the ASCII codes were covered in detail. While most text is transferred as ASCII there are other ways that text can be transferred. The popular systems are connected to word processor formats and formatted printing formats. Most of the different systems used also use ASCII as a basic encoding of the text symbols within a document but they also include different formats for layout, font and other information used to reproduce documents as close to an original as possible.

The international standard for document exchange is actually ODA (Open Document Architecture). This is little used and the standard lags behind the function of available document processing tools and software by an appreciable amount. This is unfortunate since a robust standard in this area would have enabled a multimedia extension to be used from early in the development of multimedia devices. What has actually developed in the various communities of the computing industry is a number of *ad hoc* standards that are used for document exchange, each for different purposes.

Common amongst Unix users is the use of Tex and Latex. These are tools that use ASCII text and embedded formatting commands to produce output suitable for typesetting. Also common is the use of PostScript™ as a means of transferring finished documents. PostScript is a printer language that allows documents to be formatted on paper. It is not really designed for processing by users.

In the PC world the commonest exchange format is becoming the commonly used word processing format of Microsoft's Word for Windows. In line with other products, this allows conversion to and from many of the common PC word processing formats, allowing exchange of any document in any form (more or less). It is now common to exchange documents between different platforms using non-native WP formats and be able to compile a usable document between different users. It is, however, much easier if users are using the same software and version.

The diversity of systems used by PC users is unlikely to change in the near future and so it seems likely that the compromise solution of using conversion packages is set to continue. The common factor in all of the various formulations of text document is the use of ASCII codes to represent the actual characters. Some word processors also alter the eighth bit for their own use and include the ASCII characters in the range 0 to 31 for formatting purposes.

This widespread use of the ASCII character set is biased to the English-speaking world since the variations for other national symbol sets are not as widespread, and pictographic languages are excluded. The new Unicode 16-bit representation should remove these disparities, but it will take time to upgrade systems to cope with using it.

4.3.2 Image

The coding of images for use on screen has been a feature of computer systems since they became capable of displaying high-resolution colour images. There have been a number of different algorithms used for both the coding of images and their compression. The most simple is the bitmap which is used in PCs to store images in an uncompressed format. The other forms are widely used and vary from application to application. Table 4.2 shows some of these codings and compares the storage requirements of each for the same image, that in Figure 4.2.

Table 4.2 Relative size of encoded image files

Coding scheme	Abbreviation	Image size
Bitmap	BMP	381K
Graphics Interchange Format	GIF	40K *
Tiled Image File Format	TIFF	107K *
PCX File Format	PCX	300K
Targa File Format	TGA	127K *
Encapsulated PostScript	EPS	255K *
Joint Photographic Experts Group	JPEG	19.5K

* indicates reduction to 8-bit colour

The final line in Table 4.2 is worthy of further explanation since the JPEG format is designed to be an international standard compression format for images that was developed by the Joint Photographic Experts Group (JPEG) to be the universal standard image compression method for use in a variety of future contexts, from colour fax to videotext. Furthermore the JPEG standard is likely to be the most common format in future even though it may not be at present. Not only does JPEG provide a good compression ratio but it also has widespread applicability. JPEG is now published as ISO/IEC standard 10918 — Digital compression and coding of continuous-tone still images.

The JPEG image standard is based around some complex mathematical ideas. It also contains different encoding schemes for lossy and lossless encoding of images.

For lossless encoding the image uses a predictor for each pixel that compares it with surrounding pixels and computes the difference — this difference is the quantity that is then fed into the next stage of the encoding process where the values are replaced with code values from a table — the code values are found by a statistical analysis of the original image to produce an encoding scheme known as Huffman encoding. The code table is also part of the encoded image.

Figure 4.2 An image of Nordkapp (the most northerly point in Europe)

For lossy encoding an image is considered to consist of a number of 8×8 blocks of pixels. These are then subject to a transformation to make them more amenable to coding, they are then quantized which removes some of the redundancy and picture quality, and the subsequent coding, which is similar to the lossless scheme achieves a compressed image.

These descriptions are necessarily sketchy, further details are too complex for this book, but can be found in those quoted at the end of the chapter. Compression ratios obtained by JPEG are significantly better than other coding schemes. The scheme does, however, require greater processing to produce the encoded or decoded image.

Table 4.3 gives some idea of the various compression ratios that can be obtained from JPEG compression techniques.

Table 4.3 JPEG Compression ratios

Quality	Bits per pixel
Moderate to good	0.25 - 0.5
Good to very good	0.5 - 0.75
Excellent	0.75 - 1.5
Near original quality	1.5 - 2.0

Applied to a standard image of 352×240 this would give a compressed size of between

$$\frac{352 \times 240 \times 0.25}{8} = 2640 \text{ bytes and}$$

$$\frac{352 \times 240 \times 2.0}{8} = 21\,120 \text{ bytes}$$

compared with an uncompressed size (with 16-bit colour) of

$$\frac{352 \times 240 \times 16}{8} = 168\,960 \text{ bytes}$$

Clearly JPEG offers a compression ratio that can be useful in multimedia communications if the processing overhead is not too intensive.

The other formats are mainly proprietary and are related to specific software of systems but some (e.g. GIF) are in widespread use. (See Chapter 8 for a description of images in HTML documents.)

4.3.3 Audio

Audio is one of the more problematic of media types to include in any system. This is not due to its complexity but to the diverse ways in which it has been used and

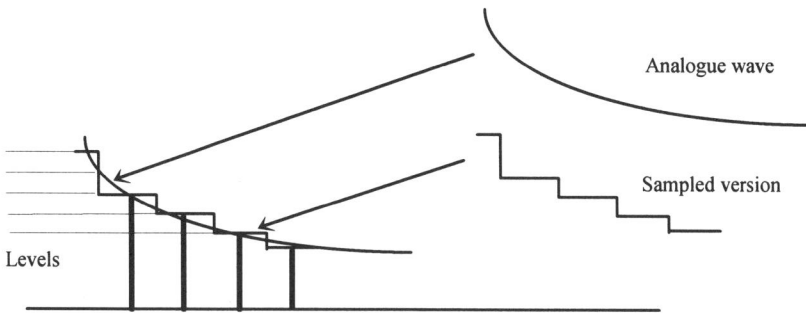

Figure 4.3 Audio wave sampling

encoded over the many years of its use in communication systems. In fact, the widespread use of voice communication systems has a number of effects on its use in multimedia communication systems.

Partly, the problem is that the use of analogue voice systems is widespread, whereas a digital encoding is necessary for multimedia computer systems. The familiar uses of audio in telephone systems, as a radio signal and in television sound are all essentially analogue voice channels. To convert these and other sources of sound requires the use of a digitisation standard. Some of these have already been briefly discussed in section 3.2.3.

The system known as Pulse Code Modulation (PCM) is the use of a straightforward digitising signal. That is, the analogue waveform is sampled and each sample is represented by a number. The number of bits used to represent the number determines the quality of the digitised sound. 8-bit sampling allows 256 levels of signal to be distinguished whereas 16-bit allows 65 536 different levels. Sampling is illustrated in Figure 4.3.

As illustrated in the diagram the use of sampling creates an error in the digital output. The use of more levels in the sampling (i.e. more bits) allows the errors to be reduced and the digital output approaches the analogue input more closely. The quality of the signal depends on the use of sufficient levels for the difference between the original and the digital output to be indistinguishable under normal use and sufficient samples to be taken per second so that the input wave is accurately represented by the digital form. The standard for CD-audio has achieved this.

CD-audio uses a sample rate of 44.1 kHz and 16-bit samples of both channels of a stereo input to provide digital audio at a data rate (as calculated in 3.2.3) of 1.4 Mbps. (The standard CD actually contains more data than this to allow for error-correction.) This can be seen as the highest quality required for multimedia use in communication. There are, however, higher quality standards used in professional environments. These will not be discussed.

For the lower quality end of the audio spectrum there are a number of different standards in use that use different sampling rates and levels and other techniques for sampling. The straightforward PCM described above is not efficient at lower sample rates so an altered version is used which encodes the differences between adjacent samples. This can be done with fewer bits per sample than PCM of the same quality. (Details can be found in the references quoted at the end of the chapter.) The scheme is called Adaptive Delta PCM or ADPCM. Using ADPCM of 8 bits per sample is equivalent to using PCM at 12 bits per sample. One standard uses ADPCM to provide digital voice in communication circuits using a sample rate of 8 kHz with 8-bit samples, this can then be transmitted over a 64 kbps connection. There are also systems which perform at lower bit rates with a corresponding reduction in quality, e.g. digital mobile telephone systems.

For the PC, standards are not a simple area. Each sound processing card originally used different standards and software and there was a large degree of incompatibility. However, widespread use of certain sound cards allowed some standards to develop, and now the 16-bit system is fairly universal with most sound cards being able to cope with simple PCM signals.

One standard that has emerged from the computer area rather than from the communication area is the use of MPEG to code audio. As described in the next section MPEG is an encoding system which is used for the compression and storage of video clips (with sound). The audio compression system can also be considered alone but is not often used as such.

The output from any compression algorithm or system will still require to be fed through a sound output device and these are now approaching a degree of standardisation as mentioned above.

In communication, however, the use of standards throughout the networks has kept diversity to a minimum and the use of ITU-T standards has allowed digital versions of communication to be carried out between different systems world-wide with common standards being the norm. Standards are available for many audio communication tasks the predominant area being voice communication, both analogue and digital. These are too numerous to cover in a book of this size. The interested reader should look into ITU-T standards such as G.711 for digital voice coding and other of the ITU-T standards.

4.3.4 Video

In digital video there are a number of standards to consider. Firstly the image standards previously considered, then the more specific moving image standards that are used for video data both in computers and communications areas. These latter are MPEG and H.261. These will be defined and explained in the following paragraphs.

The difference between video and still images is, perhaps, obvious. The video principle is that the rapid presentation of successive still images gives the impression of motion to the viewer. This requires a fair amount of similarity between successive images for there to be any possibility of the sequence being viewed as a "moving picture". This requirement also aids the encoding of the images since there is inevitably a large similarity between successive frames there will also be a high degree of redundancy contained in any sequence. Essentially, each frame differs little from the preceding one and is likely to have a high degree of similarity so this factor can be used to encode more efficiently than if the images were to be considered separate. Even in a fast moving film with a new scene every second or two there will be long sequences of frames (at 30 fps) that will be similar. 60-90 in a 2-3 second scene.

JPEG The use of JPEG cannot produce any degree of compression between successive frames of video and is not likely to allow the highest levels of compression since it ignores the main source of redundancy in video sequences. It can, however, be used to store video sequences, but will need a considerable amount of processing power. If the compression ratios illustrated in the previous section are taken as typical of JPEG the following analysis shows how much data would be generated by this encoding technique.

Example At 25 fps there are 25 images to encode per second. At a standard size of 352 by 240 pixels an 8-bit colour image would have a raw bitmapped size of 82.5 kb and a possible compressed size of around 10 kb (assuming a compression of 8:1 or 1 bit per pixel). This is 80 kbits per frame or 2 112 000 bps for 25 frames.

This is, therefore a feasible amount of data to handle but would probably be difficult for most installed communications equipment, at present, and many computers would have difficulty in processing at this speed. This does give some indication of the compression rates needed for digital video to be both viable and able to be used over communication channels. Use of JPEG in this form is called Motion JPEG and is only required if access to each individual frame is needed at high resolution or at larger sizes.

There are other considerations that can affect the actual encoding of video. Amongst these is the necessity to access the encoded data in a either a purely sequential form or in some form of random access. Also, the possibility of replay or reverse playing would need extra consideration when devising an encoding method since any reliance on temporal linearity would negate access in these modes. The specific video encoding schemes, discussed below, have taken these factors into account in their design.

MPEG Probably the main standard effort is by the MPEG (Moving Picture Experts Group). This group developed a standard at around the same time as the JPEG standard and parts of the two standards have a passing similarity. One of the main considerations of the work was to consider a data rate of 1 to 1.5 Mbps as a target since both communication systems and CD-ROM could handle data at this rate giving application areas in both storage and communication of video information.

There was also a need to consider both asymmetric and symmetric applications. Those that require more frequent decompression than compression are termed asymmetric applications, such as movies on disc or games, whereas symmetric applications are those that require equal compression and decompression time, such as video conferencing links or video mail.

The features that have been included are also an indication of the foreseen use of MPEG encoded video. These include:

1. *Random access to frames.* To allow starting of the video sequence at any point.
2. *Fast forward and reverse searches.* To view video in either direction at more than the original speed.
3. *Reverse playback.* To allow, where necessary, a reverse play mode. This is obviously not required in real-time applications such as video telephony.
4. *Audio-video synchronisation.* To maintain a close synchronisation between the audio and video signal. This is especially noticeable in lip synchronisation of talking people.
5. *Robustness to errors.* An encoding method should be able to recover from error situations and errors should not propagate through the frames or cause catastrophic failures of the video playback. This is especially helpful when using communication channels as these are never error-free.
6. *Adjustable delay time or real-time operation.* The delay between the start of the decoding process and the display of the video information is not a significant factor in operation of video playback, but is a parameter to be minimised in video telephony and other real-time applications.
7. *Editability.* To allow inclusion of other video in encoded sections. This requires the encoded unit to be relatively small and easily accessible.
8. *Flexible format.* To allow video to be displayed in different sized windows and with different frame rates.
9. *Implementable in hardware.* For maximum processing speed a dedicated chipset for decoding is desirable.

To satisfy all these requirements will lead to a trade-off between the various aspects. For example, to allow high compression ratios requires reducing the redundancy between frames, but good random access requires that as little information is carried over, from frame to frame, as possible so that all frames are accessible.

The MPEG algorithm is quite complex and only an overview is given here. The basic scheme used is based on three main picture (frame) types. I, B and P. Figure 4.4 illustrates how these relate to each other.

The I frames contain a similar amount of information to a JPEG still image. These are the basis of the encoding, they are the frames with the maximum amount of information. A P frame contains less information than an I frame and is obtained by using motion-compensated prediction from past I frames. The B frames have even

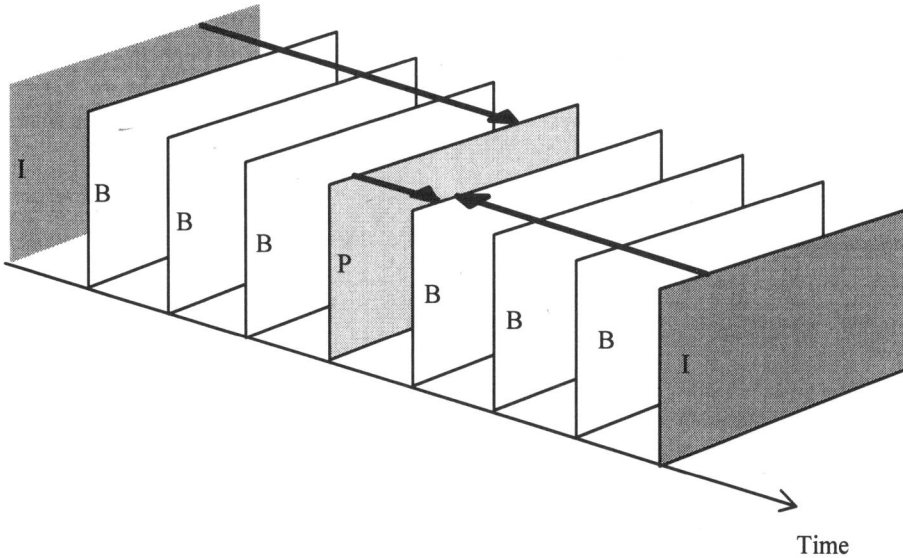

Time

Figure 4.4 Frames in MPEG

greater compression and are decoded by interpolation between an I and a P frame. So the time-dependent compression is achieved by using frames which are derived from past and future frames by either prediction or interpolation. The I frames provide convenient points of entry for random access searches and the B frames provide high compression rates. A careful choice of these parameters in any encoding will determine the applicability of the MPEG standard to a particular encoding. For example a video stream that needs high compression but little random access to frames would use a high number of B frames, whereas if editability was a key feature then the number of I or P frames would need to be increased.

MPEG audio is essentially separate from the video encoding method described above but is well-developed and able to encode a number of different audio channels for use with MPEG video. In MPEG audio, audio information is typically compressed to about 64 kbps per channel or some multiple of this to allow mono or stereo sound channels to accompany MPEG video. Synchronization is carried out via MPEG system coding a third part of the standard. However, in a multimedia system it may be that MHEG or Quicktime (described in section 4.3.5) could be used to link the different elements together to form a multimedia presentation that could be transmitted over a network or other communications channel. The finer details of MPEG are left out here but interested readers will find a good treatment in the references given at the end of the chapter.

This original version of the MPEG standard is only the start of digital video encoding. There are now plans for other MPEG standards to cover different areas of speed and compression from the lowest bit rates to multi-megabit versions.

H.261 H.261 is an ITU-T standard for video telephony. Officially called "Video Codec for Audio-visual Services at p × 64 kbit/s", it describes a real-time coding system for video signals to travel over communication systems at multiples of 64 kbps or the standard ISDN speeds. The majority of devices have initially used 2x64 kbps as a baseline for video telephony but other systems are available. H.261 is similar to MPEG but produces video at much lower bit rates and of lower quality.

The basic format for H.261 is the Quarter-Common Intermediate Format (QCIF) derived from the Common Intermediate Format (CIF). The CIF is a 352×288 frame and the QCIF is 176×144, or quarter-size. The basic layout of a frame is shown in Figure 4.5.

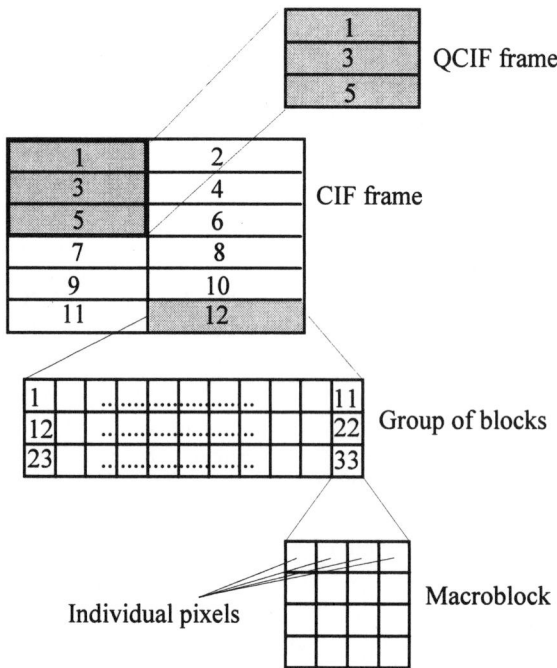

Figure 4.5 H.261 pixel grouping and block structure (CIF)

Frames can be coded as a stand-alone frame or predicted from previous frames with less information transmitted. Each frame is transmitted with a header relating it, in time, to a previous frame. So frames can be dropped and more can be predicted to allow lower bit rates to be used. The header also contains information about the size of the frame (QCIF or CIF) and other parameters followed by a group of block and macroblock data to reconstruct the frame. Each of the macroblocks is coded using a transform to reduce redundancy, as was used in JPEG.

Each pixel is coded using a luminance-chrominance model. The colour information is 24 bits per pixel, however, only the luminance component is sampled for each pixel. The chrominance component is sampled at half this rate, i.e. at 176× 144 for each frame.

Some error correction is possible using a BCH (511,493) forward error correction code specified in the standard, this corrects errors of two bits in the 511 bit code.

4.3.5 Multimedia standards

Since the interest in multimedia, as a subject, increased there has been a range of possibilities for exchanging different types of information between users and systems. It has not, until recently, been easy to standardise this activity. There are now, however, some standard architectures for the exchange of multimedia documents or multimedia sequences that can be used in communications systems and in stand-alone multimedia applications. The two methods described below come from different organisations, QuickTime being a proprietary standard and MHEG an international effort.

MHEG This is the standard devised by the Multimedia and Hypermedia Experts Group (MHEG), working under the auspices of the ISO. It defines how multimedia objects can be exchanged between systems. It does this by defining how objects can use a transfer syntax, Abstract Syntax Notation.1 (ASN.1) to be sent from user to user and it also includes the definition of the MHEG engine that uses the different multimedia objects to form a multimedia presentation. The standard also allows for interaction by the ultimate user.

MHEG allows many types of multimedia object, but supports mainly ISO defined standards, such as MPEG and H.261, JPEG and ISO6429 text. These are all encoded using the transfer syntax for transmission to another MHEG device and then replayed to the user. There is also a composite class of objects that allows a presentation to be defined in terms of a structure of objects. It provides the framework for the presentation.

The objects transferred are played out through the MHEG engine using virtual time and space co-ordinates to relate to the real space and time of the playback equipment. This is to avoid the problem of device dependent systems becoming constraints on future development.

To develop an MHEG presentation for transmission is similar to standard multimedia authoring except that the output needs to handle the ASN.1 transfer syntax used by MHEG. The output would then be sent to the replaying user's system for use. This can be done using most of the usual transfer methods such as file transfer, or email.

MHEG is fairly new and the references at the end of this chapter provide a useful starting point for further investigation.

QuickTime The QuickTime movie format standard applied originally to Apple Macintosh computers as it was an extension to the Macintosh operating system. There are now implementations available for PCs and workstations that allow QuickTime objects to be replayed on other platforms. It is, therefore, an important cross-platform standard, much the same as those discussed above. Its origins have meant that it is still more prevalent in the Macintosh world than elsewhere but it is worth describing here, since most users will come across QuickTime sooner or later.

QuickTime is more than a video compression system, such as MPEG, it is a multimedia *movie* playback system. It takes information in tracks and assembles and presents them to the user. It does incorporate video compression but that is only part of what QuickTime can do. It supports the following functions:

Video	supporting a range of compression techniques for both standard video and animation information.
	MPEG compression, decompression in software and compression with hardware support.
	Compact video compression, a lossy asymmetric compression algorithm for real-time playback.
	Video compression, a simpler algorithm, lower quality, but faster than the previous two algorithms.
	Animation compression, similar to the video compression available but optimised for animation sequences of computer-generated images.
	Photo compression, an implementation of the JPEG algorithm.
	Graphics compression, for single frames and image sequences of 8-bit images. High compression but slower decompression.
	Raw compression, allows alteration of number of bits per pixel.
Audio	using PCM with 8 to 32-bit samples and up to 65 kHz, and an audio compression algorithm that reduces information by a ratio of between 3:1 and 6:1.
Text	tracks of simple text can be used alongside the other media types.
Music	MIDI (Musical Instrument Digital Interface) is supported.
Graphics	use of Apple's PICT format is supported.
SMPTE time code	can be used for synchronization of external devices.

A QuickTime movie can combine a number of tracks of different media types into a single display entity, a movie. This contains time values placed within it to synchonise the various tracks. This allows tracks to be timed in a manner that is natural to them, e.g. if a video sequence uses 30 fps then a time scale with units of 30 per second is natural to use. The different tracks consist of any of the media types mentioned above and are combined to give a multimedia presentation. The fine details are not important here but the reader should be aware that the QuickTime format exists. Further details can be found in the book by Gibbs and Tsichritzis quoted in the references to this chapter. While there are other data models for multimedia they are not of sufficient interest for further discussion here.

4.4 OTHER MULTIMEDIA PLATFORMS

Although the emphasis in this chapter has been on the IBM-compatible PC, there are many users of larger systems using multimedia. Much of the Unix workstation market is capable of multimedia playback. There are, however, many more PC users than users of other systems in the world and much of the communications boom and the growth of the Internet is fuelled by PC users requiring access to other systems with information for use on PCs. This then does not favour the workstation user since a small market share inevitably means less effort is available for development of software and related information for these platforms.

A lot depends on the user. In many educational establishments there have been workstations used for many years to allow high-powered computing to be carried out. These are now nearer in specification to top range PCs than ever, and the differences are mainly in the operating system used and the facilities available. Users outside of education are less likely to have access to workstations and are more likely to find that the *de facto* standard of the PC is appropriate. Also, as mentioned previously, the Apple Macintosh provides a good multimedia environment, since it has been designed for this purpose with operating system extensions to permit multimedia playback and capture without too much additional hardware or software. Again the fundamental problem of these machines is the lack of a large user base when compared with the PC market.

There are some interesting developments, as have been discussed above. International standards for multimedia exchange should ease the compatibility problem and cross-platform developments, such as QuickTime players for PCs all aid the expansion of multimedia from a hardware dependent tool to a true communication medium.

4.5 DISTRIBUTED MULTIMEDIA

The distribution of multimedia objects over several computers adds problems to an already complex area. These can readily be imagined and include network traffic and bandwidth limitations, synchronisation problems, incompatibilities of end-systems, cost considerations and user-related problems. These various areas are expanded briefly in the following paragraphs.

When considering bandwidth and capacity of transmission in a distributed environment, much depends on the type of network under consideration. For example, in a LAN environment the user has a certain degree of control over the installed capacity and use of the network which should make predictions of traffic rates and data flows easier than in a WAN environment where often actual data rates are unpredictable and can fluctuate according to load and usage. This requires careful consideration when using any type of network for multimedia information exchange. The response time of the network will depend on the use by other users and the actual use proposed by the user. So, for example, transfer of live video is feasible in a LAN environment using bit rates up to 100 kbps or more which are relatively easy to achieve on a standard LAN. Over an Internet connection it can be much more difficult to achieve a minimum data rate to support this type of traffic. (There are video information exchange products and programs available for use on the Internet but they cannot guarantee connections at a certain data rate.)

Distributed environments also allow dedicated servers to be used for different media types. Indeed, it is feasible to have a dedicated server for particular types of document such as photographs in the form of a picture library or sound from a sound archive. This raises additional problems with synchronisation. If a multimedia presentation relies on information being downloaded from various sources the distribution of those sources will affect the flow of the information presented to the user. Obviously, off-line downloading would be a way to work around such problems but there may be times when an item of information is changing frequently and needs to be downloaded on use to ensure its currency. This can lead to added problems of synchronisation in that the multimedia presentation will need to ensure that the information is available at the required time. This situation needs to allow for any problems that can occur in network capacity use of popular objects and information. For example, a multimedia display may be used to give the latest news from a specific source, in audio and text. For this to be accurate and up-to-date it would need to download the information as close to display time as possible to ensure the latest information.

There may also be incompatibilities in the distributed environment. Although there are standard protocols for the exchange of information there are many different formats for the storage of information, which have grown from their use on different computers over many years. This may cause user's equipment to fail to display information downloaded. Although much effort has gone into standardisation there is still considerable variety in the end-user equipment used and this may cause various problems when information is distributed between servers and users.

The influence of cost cannot be under-estimated also. The use of distributed sources of information is often the only route of access to various specialised items. This then requires access via wide-area connections which can be expensive. Even on the Internet, where access is often charged at local rates there is still often a considerable time required on-line when downloading information from a source that is used by many users or is connected by slower data connections than is usual.

There may also be a number of user-related problems caused by the difficulty of working in a distributed environment where access to information is achieved by user clients and the underlying distributed nature of the information sources is masked from the user. This may create problems with uneven access times and other factors. The delay between attempting an access and the display of the information is a critical factor in determining the user's subjective response to an interface.

Finally, in anticipation of Chapter 8, the use of HTML (Hypertext Mark-Up Language) must be mentioned as a catalysing feature in the use of distributed sources of information. Although a special case where information sources are linked to each other to form a web of information (the World Wide Web), it is the best known of the distributed information sources in use at present. Fuller details are contained in Chapter 8 but the use of HTML documents is a useful first step towards a fully distributed information environment. However, some work on the networks used to provide access to the information is still needed to realise the ambitions of many users. That is, to have seamless access to distributed multimedia from source over high-speed reliable networks at an economic price.

4.6 AUTHORING TOOLS

Since this book is primarily about multimedia communication the need for a "director" of the information provided over a communications link or network does not disappear. While learners or researchers at a university might be using a system independently to access diverse, unconnected sources of information, this is much less likely in younger people or in more structured learning environments such as in specific training regimes. There will, therefore, be some need for authoring tools or systems to allow the structuring of multimedia information into presentations and linked displays.

Typically, these packages offer fairly similar facilities to structure multimedia presentations. Unfortunately, the lack of standards is a factor that works against their widespread use at present. Each package uses different ways of controlling the presentation which do not interwork with other packages. The development of MHEG (section 4.3.5) may go some way to resolving this problem.

At present, an authoring tool will either give a time-based metaphor where different elements are presented at times determined by the author, or allow an interactive approach where the sequence is determined by the user of the display software at run time. Both approaches require the author to supply the individual elements to the system and they are then combined into a presentation or display. The timeline approach is illustrated in Figure 4.6.

Figure 4.6 shows the display of elements in a multimedia presentation where different types of media object are shown separately. The actual placement on screen is not shown in this model and may be of importance to the author to determine the best placement of items for the correct emphasis.

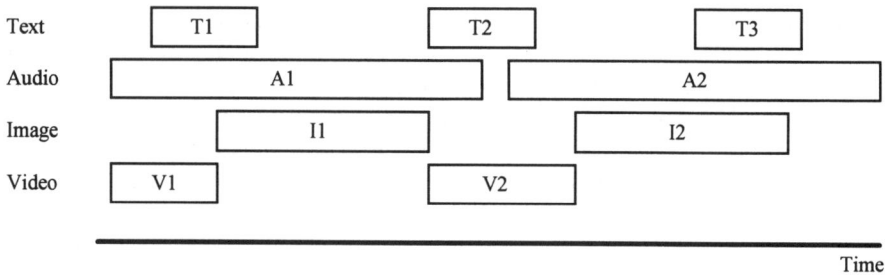

Figure 4.6 The timeline approach to multimedia presentation

Other authoring tools are based on the use of flowcharts or similar devices to show the possibility of using user decisions to drive a presentation. This allows more interactivity into the process. The use of any system which incorporates communication in the process will be more difficult since the playback of multimedia elements is not as easy to organise when the delay of networks has to be taken into account. An authoring tool for a fully-distributed multimedia presentation is unlikely to be available until the relevant standards are finalised and accepted.

4.7 APPLICATION AREAS

Examples of the use of multimedia systems can be taken from many areas. Perhaps the most obvious example is the use of distributed systems for research purposes where the use of dispersed resources is made necessary by the nature of the information required by the user. Any information from other researchers may only be available at specific sites updated by the relevant research teams. This then allows other researchers to access results from other teams as soon as they are made available rather than waiting for the formal publication in academic journals which can take some time. This relatively unstructured use of distributed multimedia may be difficult to incorporate into an authoring tool but the use of hypertext is very common in this type of educational setting where many users use it to link together items of interest around the world using the World Wide Web facilities of the Internet. This application needs little more than a hypertext editor, although it is possible to envisage a more interactive environment where the multimedia information is presented to the user without the customary delay associated with present hypertext systems and the multimedia elements are integrated into a more uniform appearance.

Another, more specific example is taken from training where the user may require access to particular information resources in order to complete a relatively restricted training task. For example, a trainee in the travel industry would need access to multimedia information about specific travel packages, destinations, carriers and tour operators. All of this information may need to be updated frequently in a live situation so the training environment would need to reflect the nature of the information used in the industry. This could be done by using a CD-ROM based training package but is probably better provided by a communications based system using real data and systems, as in a real situation.

Further examples of the use of multimedia systems are in the later chapters in this book where the different application areas of industry, education and the home are looked at in detail.

FURTHER READING

J. F. K. Buford (Ed.), *Multimedia systems*, Addison-Wesley, 1994, ISBN 0-201-53258-1.

S. J. Gibbs and D. C. Tsichritzis, *Multimedia programming: Objects, environments and frameworks*, Addison-Wesley, 1995, ISBN 0-201-42282-4.

ISO/IEC "Information technology: Coding of moving pictures and associated audio for digital storage media up to about 1.5 Mbit/s(MPEG)", IS 11172:1992.

ISO/IEC "Information technology: Coded representation of multimedia and hypermedia information objects (MHEG)", IS 13552:1993.

D. Le Gall, "MPEG: A video compression standard for multimedia applications", *Communications of the ACM*, April 1991, Vol. 34, No. 4, pages 46-58.

T. Meyer-Boudnik and W. Effelsberg, "MHEG explained", *IEEE Multimedia*, ISSN 1070-986X, Vol. 2, No. 1, Spring 1995, pages 26-38.

D. Pan, "A tutorial on MPEG/Audio compression", *IEEE Multimedia*, ISSN 1070-986X, Vol. 2, No. 2, Summer 1995, pages 60-74.

EXERCISES

1. What standards are relevant to the applications described in section 4.7?

2. A LAN has a number of users connected simultaneously with a 10 Mbps CSMA/CD bus system. If the utilisation is best kept to 50 % (to guarantee response) and there is a 10% protocol overhead, calculate the maximum number of simultaneous users of 100 kbps video telephony applications able to use the LAN before the limit is reached.

3. A multimedia presentation consists of 10 minutes of video and audio information with 10 pages (each page is a single screen, size of 80 x 25 characters) of text and 25 photographic images (each is sized to 640 × 480 pixels)

    Calculate the size of the whole presentation if the authored script requires 50 kbytes, text is plain ASCII and the following are used.

    a) JPEG images, MPEG video and PCM sound of 8 kHz sample rate and 8-bit stereo samples.
    b) Bitmap images, Motion JPEG video and high quality audio at 44.1 kHz sample rate, 16-bit stereo samples.

4. An authoring station requires access to uncompressed video and sound to allow editing and cutting before compression. The output from the process is to be distributed on CD-ROM. Calculate the likely disc storage requirements for the system to be adequate for this purpose. State any assumptions made.

5

MULTIMEDIA IN BUSINESS COMMUNICATION

Summary: How computers and communications integrate information processing, storage and transmission using different media types in a business context. Detailed examples to show how multimedia communication can be used in an organisational setting. Information and business organisation for the "global village". Teleworking and the implications for its development using multimedia communication.

5.1 BUSINESS ORGANISATION AND COMMUNICATION

A business is a complex entity which is basically an organisation of people working together to provide goods or services in exchange for money. This involves making goods or providing some service to a customer and at the same time being a customer of a supplier of goods and services. So every business can be seen to be in at least one supply chain formed from a supplier themselves and a customer, as shown in Figure 5.1.

The typical scenario is that some materials are supplied to a manufacturer who makes a small part of something (a component), these are then combined into assemblies by other manufacturers. All the different manufacturers who make parts supply them to a product manufacturer who puts them together. Other companies supply goods and services such as machines which assemble parts or make individual components. Other companies provide the machines and equipment necessary to do the paperwork involved in trading between companies. Eventually the goods are purchased by a consumer and the chain ends. So from raw material to finished product a number of different companies interact and co-exist to form supply chains and provide the consumer with goods or services.

End customer

Consumer product

Company A — Customer

Assemblies

Company B — Supplier to Company A

Components

Company C — Supplier to Company B

Raw materials

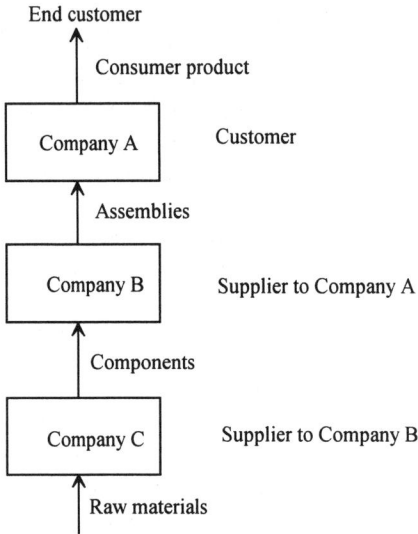

Figure 5.1 A supply chain

A more realistic set of links between companies is shown in Figure 5.2. In this figure a supply network is shown where a number of suppliers feed parts, components and other goods and services into a number of different stages in the product life cycle. For example, in vehicle manufacturing a steel supplier may supply sheet steel to panel manufacturers and steel blocks to engine manufacturers. The latter being a more complex process that requires more stages and more suppliers in the chain than the

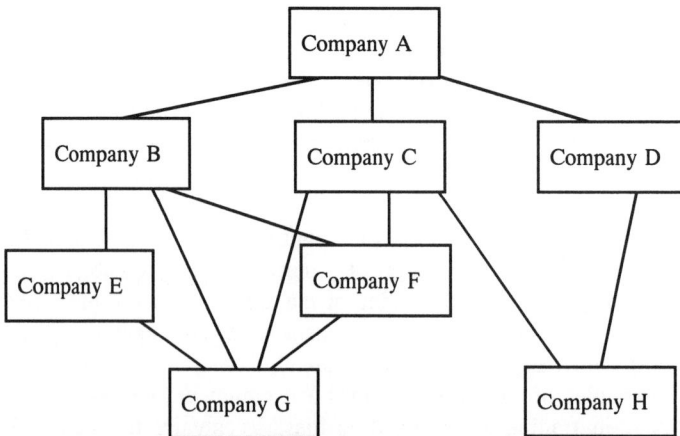

Figure 5.2 A supply network

making of body panels. Also a supplier of general services may interact with many different companies in the chains and have input at all the different supplier tiers.

So the world-wide business community forms an interlinked network of companies and concerns that are required to co-exist to continue in business. This co-existence requires a complex interaction between different companies, and inside individual companies that is based on communication between people and between people and computers.

To interact with any business requires that the infrastructure be in place. Even the smallest business now needs a telephone to communicate with suppliers and customers, but often the requirements of business are much more complex than can be supported with just a telephone or voice communication. Therefore, multimedia communication technology is increasingly seen as the way to better communication between companies and between people in the same organisation. If the situation is analysed it will be clear that many different types of communication take place between the various people involved in a business and this communication takes many forms, from the spoken word to complex documentation with images and text. Figure 5.3 shows a simplified version of the communications that take place within a company when a goods order is received. The figure shows the flow of information around the various departments in a company, assuming that text is used to convey the information, which is often the case. The text being a paper document. Also there are other forms of communication used between people when communicating that are not shown e.g. non-verbal communication such as body language and gestures. These may be important and could be conveyed if a visual medium were used to communicate.

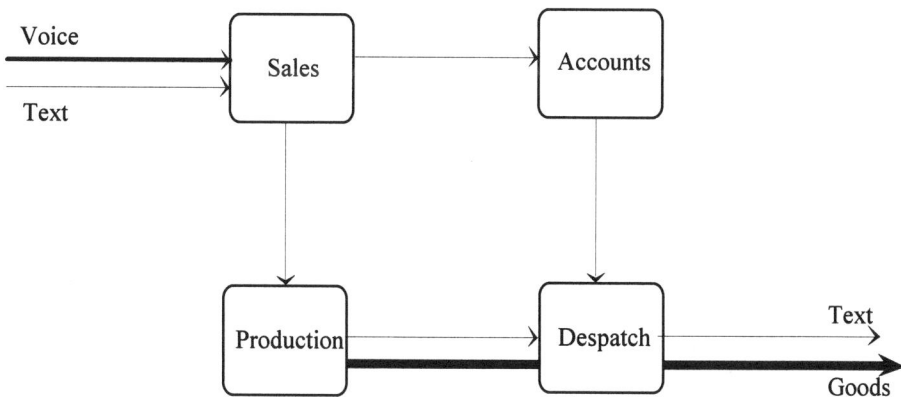

Figure 5.3 Internal company communication

What is also clear is that the future of business and industry will be changed by the introduction of multimedia communications technology. How it will be changed is open to speculation and is only briefly discussed at the end of this chapter. However, the agents for change are already here. The ability to communicate over any distance from one company to another and to send machine-readable documents between systems are all part of the multimedia "revolution" that is changing the way business is carried out. The scenario outlined in Figure 5.3 will not be changed by the introduction of this technology but will be speeded up, made more efficient and allow faster response to customers.

Furthermore, individuals can now also benefit from the technological advance. The ability to communicate between employees and their place of work has been used for many years to keep information up-to-date and make best use of employees' time. Now, the number of possibilities is increasing, with the availability of faster communications links and better mobile systems providing individual users with better access to corporate systems than ever before.

So the combination of faster communication links and better information technology has allowed the changes to take place in the business world. This has not radically altered the processes of business at present, but with the spread of multimedia communication technology, the changes are inevitable.

In the following sections the actual communication requirements of business will be considered. This will, inevitably, cover a lot of the processes of business which may be familiar to some readers and not to others. However, it is hoped that the treatment will be sufficiently interesting for it to prove useful to even the most experienced business user. For those not familiar with the ideas contained here it should also prove to be of interest.

The main aim of this chapter is to cover the most significant areas of businesses that require communication to function. Therefore, the use of devices such as the telephone will be covered as will the more advanced topics that are only now becoming possible on any scale. Firstly a short discussion of the nature of businesses is necessary, followed by a look at the use of communication in business, before discussing the role of multimedia in business communication and the prospects for the future.

5.1.1 Business sizes and types

The range of businesses that exist starts at the single individual who is often self-employed to the huge multinational corporation. Some are very specific and provide a specialised range of goods and services while some cover many different areas of business activity. The diversity in business is masked, to a large extent, by the use of common definitions and terms used to describe roles and functions within each organisation. Many different organisations are grouped together as "business" when the only common feature is that they transact business between themselves and customers and suppliers. This is, therefore, the common point for all businesses. They all exist to supply a perceived need (although there are times when the need has to be communicated to the customer!). The supply could be of goods or services, but will eventually involve the payment for the supplied items.

When the need to exchange money occurs other systems can then take over. Much of the financial world (which is a subset of the business world as a whole) is highly computerised. It is, therefore, relatively easy to transact business without the need for (physical) money to change hands.

It is worth noting here that the origin of money was as a token for exchange and, as such, only represented the worth of the goods or services it was used to purchase. There are now, however, complex systems of exchange which have given money a value of its own, in a sense, and this value is now used to judge the worth of everything else!

The financial sector can be viewed as a special case since the majority of "products" produced or handled in this sector are services and this allows much more use of information technology to perform regular business transactions. A typical example of this is the use of computerised payroll systems, where an employee in some organisation is paid via bank transfer and the money they are paid is transferred from one account (the employer's) to another (the employee's) without being turned into "real" money. Often, it also disappears from the receiving account by a similar process and is not turned into actual cash. This use of information technology to communicate "money" is only one way in which businesses can benefit from improved communication using multimedia. A number of others will be outlined in the following sections.

In addition to the various sizes of business concern that are common there are also a number of types of business. Most of these are distinct, and similar businesses tend to operate in similar ways. For example, there are manufacturing companies that follow the model outlined above, they buy parts and produce products to sell to customers. Most of their efforts are put into running a business to support the production process. At the other end of the spectrum there are the purely service industries that do not produce goods but provide a service to a customer, for example an office cleaning contractor is often employed to clean office space but does not have a direct input to the production process.

5.1.2 Use of communication in business

These differing types of business all use communication to drive the process of business. Without effective communication many businesses would not be able to function at all and with some malfunction of the process the business could quite quickly be in serious difficulty. Consider what would happen if all the telephones suddenly stopped working in an insurance brokers office. No new business would be able to be started, the office would have to rely on letters from customers to transact business requiring an extra few days to complete any paperwork, assuming nothing went wrong in which case in could take much longer! This is, however, how business was transacted before the invention of the telephone. It would now be considered strange to try to conduct business without one.

Newer technology has not yet had such an impact on business as the telephone has had, but most of the new communications technology has only been in existence for a few years whereas the telephone has been around for over 100 years. However, the use of computers in business has had a noticeable impact and it would now be difficult for many companies to exist without access to information held on computer. It is only a short step between the use of computers to store business information and the combination of computers and communications to actually transact business on-line and to conduct much of the companies communications using the medium of the computer. In the following sections the various types of communication used in business will be described and discussed and the impact of communication technology and specifically multimedia communication technology will be assessed.

5.2 BUSINESS COMMUNICATION

Inside every business the process of communication takes place every day transferring information from person to person in order to smooth the running of the business. Machine operators need to know what articles to produce, office clerks need to know where to send invoices, directors need to know how well the company is performing, production managers need to know if jobs are being done on schedule and so on. A complex web of communication takes place all the time between the individuals in the business, and between them and others outside the business, both customers and other agencies.

5.2.1 Internal and external communication

To provide a complete list of the communications that take place in any company would be difficult especially as some companies have extra forms of communication that may not occur in the majority. Therefore, the following list of means and types of communication should be taken as typical and incomplete. It does, however, indicate the range of communications that do take place both internally and externally to any business.

Internal communication. Inside a company the types of communication that are used can generally be confined to two main types, inter-departmental i.e. between departments and intra-departmental i.e. within a department. Added to this the communications can be divided into layers where the more senior managers are at the top of the levels and the various employees at their respective hierarchical levels below. Communication can then be termed as being vertical (up or down the levels), horizontal (between members of the same level) or diagonal (inter-departmental and vertical at the same time).

Some examples of inter-departmental communications are:

- memoranda — notices of meetings, requests for services, internal transfer documents
- service documentation e.g. job schedules, progress reports, delivery notes, picking lists
- sales figures and projections
- promotional literature
- company newspapers.

Similarly, some examples of intra-departmental communications could include the following:

- memoranda
- voice conversations
- administrative data e.g. request for holidays, staff sickness forms, etc.
- departmental planning documents.

Clearly, some communications can start as departmental communications but soon become inter-departmental so the dividing line should not be considered as rigid but flexible to accommodate a change in use. The range of communications is quite large and many of these can use different methods of transfer both using traditional media and electronic types, but the communications will still have to take place. A business depends on its internal communications to function effectively and replacing one medium by another should not alter the communications requirements of the task. It may, however, have other less immediately noticeable effects such as reducing staff morale or alienation of managers.

External communication. The communication that takes place between one company and another and between companies and other agencies is generally more formal and restricted since it is usually connected to the actual business that is taking place between the organisations. These two instances of communication can be considered separately as two different categories. Often the inter-company communication is concerned with the trade of goods or services and this will involve communication from the design and planning stage of production.

Examples of inter-company communication include:

- orders, invoices and other trade documentation
- drawings of components and parts
- work schedules
- advertising material

and between a company and other agencies:

- information for government regulations
- standards information
- safety information

These brief examples show how varied is the information that is transferred between companies and other bodies. Most of the information that is transferred has traditionally been text and voice information with image or graphics being used in special circumstances, such as drawing information and photographic material for advertising. However, the use of paper has been a common factor and the modern business has, until recently, been totally reliant on an efficient machine for moving paper around and between companies. With the growth in the use of computers and the need to reduce paper usage for environmental considerations, the emphasis is now on efficient use of electronic communication to achieve the same ends as in the more traditional approach, and to satisfy the needs of the business. This is now technically feasible with multimedia communication technology and the next section looks at the typical technology that is found in the business environment and its use in aiding communication and supporting the business.

5.2.2 Technology for business communication

The typical business does not have a high degree of computerisation, although this may change in the next few years. Many businesses that are outside the computing industry often have very few computers and use them for relatively straightforward tasks such as accounts and payroll preparation. In the communications area most companies would find it difficult to work without a telephone and these are seen as an infrastructure device which will be part of the necessary technology for conducting business. Also in the last few years the fax machine has become nearly as universal as the telephone and is now the most used method of transferring documents at speed between companies. (There are still carriers willing to do the same process with paper as often a faxed document does not have sufficient quality or legal standing.) There are some businesses that have taken the next step forward towards the integration of computers and communication technology and installed electronic mail for use between its employees and other companies, but they are still a minority of businesses. In other areas, such as video conferencing, there are even fewer users.

This piecemeal implementation of different technologies for communication has been carried out over the last hundred years or so and has slowly supplemented the use of paper exchange and face-to-face voice communication to transact business. It has not yet supplanted it, although great inroads are being made in some industry sectors, especially those that rely most heavily on EDI (Electronic Data Interchange), or Electronic Commerce. There are still, however, many companies where the communication infrastructure relies on a multitude of different devices, some of them connected to computers, some standalone. The currently available technologies, described in Chapter 2, are able to accomplish an integration of communication so that computers can be linked together to exchange information, people can use the technology to communicate with both voice and image and documents can be exchanged in electronic formats, almost instantly.

This integration of communication and computers has been promoted as the "information technology revolution" for many years and is still to happen for many businesses. It is not that they are reticent about the use of technology, but that there is a currently useful system that allows business to be transacted between them and other companies. While that paper-based system is able to function the upgrade to an initially more expensive system of communication is seen as a drain on resources. However, in a modern business new technology will eventually make a mark and the use of integrated computer and communication technology will be commonplace. The first step to gaining greater acceptance of new communications technology will often be the need to use a particular piece of technology such as a video telephone, or the need to install a system such as EDI. These upgrades to current business practices often come about by supplier and customer demand. For example, the ability to trade via EDI has been a prerequisite for many suppliers in the automotive supply industry for a number of years and it is now unusual to find many automotive suppliers without some form of EDI. This is not to say that they all have EDI integrated with their company computer and communication systems, but they have responded to the need by gaining a foothold into EDI. After time the benefits of closer integration with other systems in the company become more obvious.

Mostly the ability to communicate requires only a computer and a communications link. The scope of the communication will be limited by the performance of each of these components. A fast, powerful computer will be able to take advantage of any communication link up to quite high speeds. These are often shared, as illustrated in Figure 5.4, and the exchange of information between companies is often carried out by a number of different computers connected to LANs and communicating via a common outward communications line. The speed of the connection is then not the only consideration when investigating the communication needs of a company.

There are, as has been seen earlier in the book, some applications that require guaranteed capacity to function effectively, e.g. video telephony. If these are used then the ability to set aside the required capacity is crucial, otherwise the use of the facility will be degraded and the other applications using the communications facilities will also suffer. So, careful planning needs to be done before installation of any communications equipment so that its use and effect are known. Some systems such as ISDN can be installed to give both an aggregated capacity over all the lines used and individual guaranteed capacity links.

External communications link

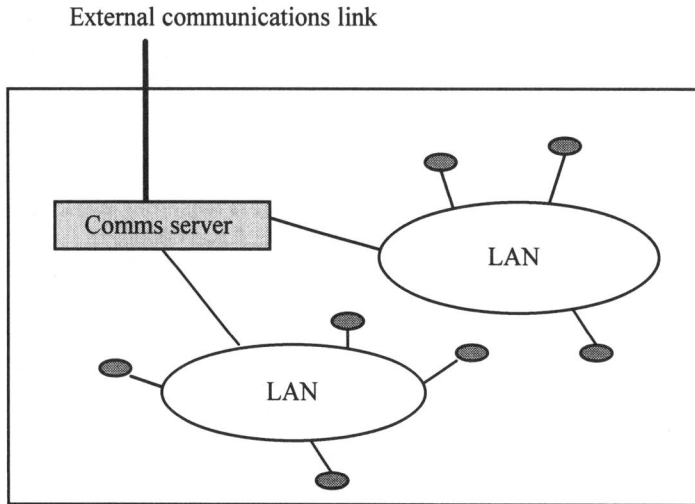

Figure 5.4 LAN and communications set-up

5.2.3 Examples

The following examples show how technology has become more integrated with the use of computers and communications links. The examples show the evolution of the communication process for various typical exchanges that are used in business everyday.

Personal contact has evolved from the *face-to-face meeting* where it was necessary to travel to conduct business, to the use of the *telephone* where it is only necessary to be able to dial the correct number and for the other person to be available. To allow more flexibility in answering the telephone the use of *answering machines* and *voice mail* was developed along with *call diversion* and *mobile telephony*. Finally to allow a face-to-face meeting at a distance the *video telephone* was developed. This further allows *video conferencing* between a number of users. These facilities are now all available as part of most communications offerings from any service provider.

Document exchange has been the standard way of conducting business for centuries. So much so that the written document occupies a relatively unique place in the law of many countries, forming the basis of many systems of legal redress and contract law. The paper document has been exchanged by carriers or couriers which have evolved into *postal services* in most countries. Until the 1980s this was the only way to exchange most documents between companies. Short messages could be exchanged via *telex*, but this only deals with text exchange. In the 1980s the *fax* machine became common with the spread of the Group 3 fax standard, and until *multimedia electronic mail* becomes more common, the fax machine is likely to be the mainstay of most businesses' rapid document exchange. The postal service is still the most likely route for a non-urgent transfer of a document.

Trading The business of trading has now become the exchange of formal documents that accompany the goods or services that are being traded. So the purchase of goods is now accompanied by the necessary *orders forms and invoices* being sent from customer to supplier and back. This has been replaced in some industries by the large-scale use of *EDI* or *electronic trading* where the formal documentation is done via computer links and the paper copy is not seen. This could further evolve into a complete *electronic commerce* system where every contact with another company could be achieved on-line from browsing their catalogue to sending the final payment.

A brief summary of some of the examples is given in Table 5.1.

Table 5.1 Computer and communications evolution

Human communication	Stand-alone device	Enhancements	Computer communications integration
Voice	Telephone	Answerphone	Voice-mail
Text (Postal service)	Fax	Fax polling	Multimedia email

These examples have shown how communications have changed over the last few centuries and developed into the modern business communications environment. This is fast becoming the accepted method of doing business. Most of the technological advances have been small and their impact has been equally small but, when taken as a whole, they add up to a new method of transacting business that is both faster and more personal. However, there are still many businesses which do not see these advances as having any importance and are reliant on the traditional paper-based communication methods to do business.

Table 5.2 summarises the technology that can used in business and the communication purpose that it serves.

The relationship between communications and computers is shown in the table and these are the fundamental technological options available to the modern business. The combination of communications and computers enables the rest of the technological options in the table and allows multimedia communication to be used for many of the communication tasks that are required in business. The next section looks in detail at the role of the various media types that are used.

Table 5.2 Business technology options

Technology	Application	Underlying technologies
Telephone	Voice comms	
Fax	Text/image comms	Telephone
Email	Text/image comms	Computer and networks
Video telephony	Video/image comms	Computer and ISDN
Electronic Data Interchange	Business document exchange	Computer and networks

5.3 ROLE OF MEDIA TYPES IN BUSINESS COMMUNICATIONS

The use of different forms of communication is crucial to the smooth running of any business; telephone, fax and the postal service form the basis of most communications that are used. Increasingly the influence of computer networks can also be seen in businesses with EDI and other transfers of information between companies. This section will describe the different ways in which the various media types are used in business and show how they are all used to convey vital information for the function of business, followed by some examples of the use of different media in business.

5.3.1 Text

Text is a fundamental type of information used in business. Much of business would be unable to function effectively without the written word. The use of standard contracts and letters, the need to "have it in writing" is an elementary part of the running of any business. Text has, therefore, been the basis of business for centuries. This reliance has two effects. Firstly it formalises the business process so that it is relatively simple to transact business by maintaining the use of text as the medium of exchange of information; secondly, it produces a reliance on the status quo that is often a problem to new and innovative processes. These can be seen as positive and negative aspects of the use of text. The use of written text on paper has entrenched its position in law and in modern business culture as the foremost means of forming agreements etc. between companies and between customers and suppliers.

This has a significant effect on multimedia communications technology and its take-up in business. The use of paper as the recognised means of text transfer to the virtual exclusion of other means delays the introduction of electronic means of communication between organisations. There are many instances of the use of text which delays the use of speedier technology. For example, the increase in telephone banking has had a marked effect on the way individual customers can do business with a bank. Many transactions can be done solely over the telephone, but the insistence on signed text documents often delays some actions by a number of days since they are sent by the postal service and require signature.

Previously, in Chapter 1, the example of an order was outlined briefly. The example showed a text based order which is similar to part of many that are used in business and industry every day, as follows.

Order

Part Number	Quantity
234556671	25
23456673	100
23456688	10

This would, however, be part of a much larger document which could look like the more complete example in Figure 5.5. Here, the order has been included with the supplier's name on the official order stationery of the company. This would be similar to many of the orders that are sent out in industry every day.

There would also be other "standard" information included on the form, possibly on the reverse of the order, such as legal requirements and restrictions, guarantees and other information pertaining to the transaction which is the subject of the order. There would also be a corresponding invoice and despatch note sent out when the order is fulfilled by the supplier, this would have a similar set of "standard" information included or attached to the document. As has been noted already, much of business is formally transacted in this way but it can also be done via electronic communications links.

The use of text is, however, much more widespread than this simple example shows. Most internal documents and notes are also based on text and the whole of business is currently geared to handling the large amounts of paper documents that are created (and ultimately destroyed) by the process of business. There are individual companies who specialise in the reproduction of paper documents and do little else quite often. To change this reliance on paper as the means of transferring information will inevitably take some time as this is more of a cultural process than a technological process. Indeed, the information can easily remain the same in an electronic transfer as in a paper transfer and even be accompanied by more information than is used on paper, but there is still often a reluctance to rely on a paperless system altogether. Obvious examples of the opposite are in the computing industry where not to use the technology would seem to invalidate the company's reason for existence! However, in general most businesses are still paper-based for most external transactions.

Hightown Engineering Ltd
Newbridge Industrial Estate
Hightown
Tel: 0119 234 3455
Fax: 0119 234 3456

To: Hightown Component Services Ltd,
        Bridge St,
        Hightown.

Official Order

Part Number	Description	Quantity	Unit Price
23456671	Widget	25	0.04
23456673	Grommet	100	0.20
23456688	Flange cover	10	2.30
		SUB-TOTAL	44.00
		VAT @17.5%	7.70
		TOTAL	51.70

Figure 5.5 A typical order form

One of the main areas of the use of electronic communications technology is in manufacturing industries such as the vehicle construction industry where the use of EDI (Electronic Data Interchange) has been popular for many years. This has been brought about by the use of trading agreements between customers (vehicle manufacturers) and suppliers (component manufacturers). These set out the rules of trade between the partners and allow the use of electronic documents to be sent via communications lines rather than a paper version to be sent via the postal service.

Text information will continue to be a main medium for communication between businesses and for the foreseeable future the simple communication of information and the relatively unambiguous nature of text makes it ideal as a means of formal communication between companies. This allows business to be transacted on a sound basis. The transfer of the communication of text from paper to electronic means may have some initial problems but eventually much of the current information that is transferred by paper will be moved by computer. This will have some effect on both the companies involved and the wider community as the reduction in paper used and printed will have knock on effects on the environment, paper providers, paper service providers and many other users of these systems. Other media types will also have the same effect but probably to a lesser degree.

5.3.2 Structured information

The use of structured information in business has always been a part of the fundamental structure of a business's operation, but it was not until the use of computerised databases that the ease of use led to a large growth in the use of structured information. The database has fundamentally altered the scope of business. The computer has allowed access to, and use of, large amounts of information that can be sorted, sifted, grouped and individually used for the benefit of the business. A typical database in a company might be the names, address and account information of customers. This would be viewed as a fundamental information source for the business. A database such as this allows new product information to be sent to previous purchasers and other users of related products. For example, if a company launches a product such as a new computer printer then it could send advertising material to customers who have bought similar printers in the past, especially if the expected life of such a printer is near its end.

This type of information is crucial to targeting customers for direct marketing of products. It may not be used in such a specific way as outlined above, but a broader marketing campaign may use databases of customer information based on market research questionnaires. For example, many marketing exercises rely on directing mail to particular groups of customers. Often this is based on information that indicates that they are the customers most likely to buy a particular product. This is based on information, often supplied by the customer, such as the questionnaires used in free prize draws or as requests for product information. The databases are queried to provide lists of customers to be targeted often based on information that does not directly link them to the product but which uses personal information such as lifestyle profiles which are supposed to indicate a customer is "likely to buy" a particular product.

Databases of structured information are, however, much more than just customer lists. There are many areas of business where structured information is used. Product information, specification documents, product tracking, employee lists, accounting information and much more are all possible subjects for storage as structured information.

The communication of structured information is, therefore, of great interest to the business person. Although some small businesses may only use a rudimentary database of information most businesses use structured information to some extent to promote their products, or just to keep track of their employees. There is often a need to communicate this information. This could be internally in the business or externally to other agencies. For example, the use of BACS (Banks Automated Clearing System) where details of payments are sent direct to a bank for subsequent processing is now commonplace in UK industry and elsewhere. This type of service relies on the use of structured information being transferred from computer to computer over a network or telephone connection.

Structured information held on a database can be any size. Small companies may hold small databases but much will depend on the type of business involved. Some companies only trade with a small number of customers whereas some trade with many members of the general public. So, structured information on databases can range from the small to the very large and all this variety may need to be communicated at some time. Obviously, there may be times when partial information from a database may be communicated and others when it is necessary to send the whole database. Mostly structured information will be fairly small in volume and any of the options for communication will be applicable but, if there is a need to communicate whole databases of information, which can be many megabytes in size then careful consideration of the communication options is needed. For example, consider a company that stores information about 10 000 customers, each of which has a record containing, name, address, personal data and other information related to product purchase, each record being 400 characters long. The whole database will occupy 4 megabytes. To back up this database over a communications connection will take the following times at the quoted speeds:

Modem speeds	9600 bps	55 min 33 s
	14 400 bps	37 min 02 s
	28 800 bps	18 min 31 s
ISDN	64 k bps	8 min 20 s
	128 k bps	4 min 10 s

This shows the value of the newer technology for this type of information interchange. If this type of information is regularly sent the investment in ISDN would be useful. The following scenario is an example of this type of information transfer in a multimedia context.

Example A company wishes to attract customers to its products without setting up expensive shops in shopping malls, centres and arcades. To do this it uses multimedia kiosk terminals to display information about its current range of products on the kiosk screen. This information is contained in a product database of the company's 450 products, each of which is manufactured by other companies. The database includes video and image material for some products and text and structured information about all of them.

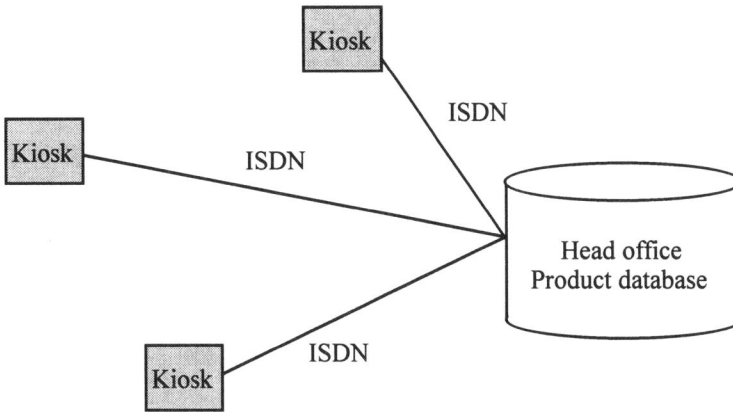

Figure 5.6 Multimedia kiosk product database communication

The kiosk communicates to the head office by means of ISDN lines and the data is updated daily to change current options, when new items come into stock or are deleted from the current range, and when special offers are to be promoted. The software for the kiosk is also downloaded to the kiosk over the ISDN connection. The maximum capacity of the kiosk is 500 Mb, the size of the hard disc, but it is rare that the whole disc is updated. ISDN is the only suitable medium for this application since it is easily accessible in the environment of the kiosk and the only fast option for communication.

Individual customers can communicate their details to the company by keying in data on the touch screen and place orders for home delivery in the same way. This kiosk system is illustrated in Figure 5.6. Although the example in Figure 5.6 is primarily about the database of company information stored on the kiosk application it also contains information stored in other forms such as video and image. This will be an increasingly likely option as the trend to better quality of information continues.

5.3.3 Image

Image information has long been used by companies to communicate information especially about products and services. A company often produces a catalogue of information about its products and each item may be illustrated in the catalogue to entice potential buyers. This may be in the retail industry where there has been a growth in the numbers of "catalogue shops" in recent years, or in a more industrial setting where component suppliers often produce catalogues of standard components for prospective users.

Image information also has its own industry based around photographs, graphic art and other tools of the image maker. This is a special case of what the rest of industry will do with image information but is often a good example of what can be done with the communication of image information.

Many companies use images to enhance trade and much research has been done over the years to indicate that images enhance the prospects of sale in all areas, when compared with text only. Consider the following text from a retail catalogue:

3 LIGHT FITTING.
Brass effect 3 light fitting
with white glass dish
shades. Height 25.5 cm,
spread 46 cm. Max 60 watt.
3 x golf ball bulb BC. Cat.
No 123/2345.

This actually describes the object for sale quite well but does little to impress without the picture of the item for sale. The image of the light fitting is shown in Figure 5.7. This illustrates the advantage of an image in saving the amount of text needed to describe an object, which in this case would probably still lead to a completely different image being imagined by any two different readers of the text!

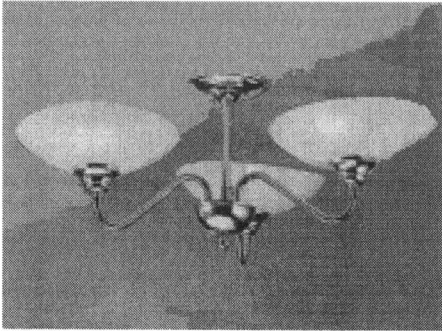

Figure 5.7 An image of a light fitting

There is, however, more value in the text than has been admitted so far. The technical details have been included with the description so that buyers know the size of the object, the amount of light it could be expected to provide since a maximum power figure is given, and the type of bulb it uses. These are all details that are difficult, if not impossible to show in an image.

There is, therefore, value in a multimedia approach to information about objects such as goods for sale which can be described by different means using different media types. Image has long been a useful tool in the sales armoury but it must not be forgotten that there is often more value in simple text descriptions where they are appropriate. Any reader who looks through a catalogue will most often find that a combination of text and image is the frequently preferred option.

Much of the growth in the number of companies using the Internet has been in the area of promoting their products. This has largely been done on the World Wide Web and primarily this is via the use of text and image. (Although any media type is possible, the majority of use is still text and image.) This is mainly a new working of traditional patterns of selling and advertising which has used the image and text approach for a long time. Now, the hypertext documents allow users to browse much easier and interact with the catalogue to a degree using database query techniques to focus on particular items or services that are available.

There are also many other areas of business in which image information is important. Examples include:

- Engineering drawings
- Product design prototypes
- Company newspapers
- Employee identity cards etc.

5.3.4 Audio

The use of sound to communicate is widespread if voice communication is included. It accounts for much of the communication that takes place in and between businesses and will continue to do so. Multimedia communications technology will only enhance the ability to communicate with audio especially voice. This is mainly because voice communication has been a fundamental human communication system for many centuries and is part of human culture and is likely to remain so for the foreseeable future. Voice communication is not the only use of audio channels.

The other uses of audio are often more specialised, with high quality sound being a product in its own right. The sale of musical recordings on compact disc is a useful case of this specialised use of audio. The ability to encode music digitally and transmit it over communications channels will allow access to a wider range of products than when the same digital information is provided in CD form. Users can download music from servers specialising in audio information and playback the music at leisure. If the communication channels have a high bandwidth it would be possible to playback on-line However, most channels in use at present are not able to do this.

In a typical business the everyday communication medium is via sound and most employers and employees will need to communicate with each other to be able to achieve tasks and co-ordinate production and so on. This has always been possible using face-to-face voice communication and then the telephone, and voice mail.

Increasingly, voice is seen as one medium of many and as a supplementary channel of communication.

For example, in a business presentation, the presenter will often talk for most of the time the presentation is taking place and also show text and image to reinforce the points made in the talk. The audio component is often the main intended means of communication but the text and image are used to add extra information that is difficult to convey using voice alone.

The use of voice communication is so widespread it is difficult to find examples where it is not used in business. It is the fundamental human communication tool and was one of the first types of information to be communicated over a distance using electronic communication technology, i.e. the telephone. Now it is possible to communicate all forms of information, as has been indicated and discussed here, it is easy to forget how fundamental the use of voice communication is to the communication needs of humans. Most of the more complex forms of communication include voice or sound as an integral part of the communication process. This is certainly the case with the use of video where the audio soundtrack is usually taken as being inseparable from the video pictures.

5.3.5 Video

The final type of information is video. As has been noted earlier in the book, video is the most intensive of media to communicate. It requires more data to be sent per second of video information than any of the other types discussed. The use of video information is a relatively modern packaging of information. The ability to produce moving picture is only about 100 years old and the electronic form of this, video, is much newer. The use of the medium has rapidly developed since the time it became possible and the use of video is now widespread with many amateur users of the medium in evidence.

In business, video is often seen as an expensive medium. Which is often true if a fully professional promotional video is required. It can, however, make a more immediate impact than any other form of communication. Especially when combined with audio, text and separate images. (It can also be quite the opposite if made badly!) There are also situations in which the use of video can make a lengthy voice communication much shorter. For example if an engineer is trying to solve a problem when at a distance it can often be simpler to point a camera at the problem than to try to discuss it using voice. This is especially true of equipment failure when a video of the problem allows the engineer to find the information needed without asking a lot of specific questions about the problem. this is not true in all cases but can often save much time and travelling.

In its simplest form video is a moving image that can be used to convey information. A typical use is video telephones where the two persons communicating can see each other over a distance. This does not actually convey much more information than in a simple audio call except for people who haven't met prior to the communication. It can communicate some facial expression and body language but the field of view and the resolution of the technology will need to be good enough if this is a requirement as many video telephones are designed to perform to a low bandwidth standard not a particular information communications requirement.

Video has probably not reached its full potential as a business tool as yet. There are many situations where the use of video would probably save time and money but the technology has not always been available to allow it. Particularly the case of remote video conferencing is seen as a principle use of video telecommunication in the years ahead. This allows users to meet as if in a single room and discuss ideas and proposals. There are differences between a real meeting and a video conference that still need some research to allow best use to be made of the medium, but the technological opportunity is available if business want to use it. Increasingly, video conferencing is becoming a lower bandwidth option that is getting down to the desktop. Video conferencing is now an option that requires serious consideration.

Furthermore there are many other applications of video that could be used in business. As database servers become more powerful and able to provide video to users on a demand-driven basis the video archive or video catalogue will become more common and users will be able to browse through company databases of products where their characteristics can be displayed more fully than is possible with image or other media types. An example would be a video to illustrate the ease of use of some equipment such as a video recorder or the performance of a car.

5.3.6 Typical examples of business information

Multimedia communications technology now makes it possible to use both image and text along with sound and video to sell goods. This is especially useful for items that rely on the way they move or sound to promote their sale. A good example is a film. Most cinema films in the past have relied on advertising in two forms, parts of the film itself (a trailer) can be used to promote the film or a traditional approach used outside cinemas in the past was to put images from the film on display (film is only any use as a promotion medium where it can be shown, e.g. inside the cinema). Both often included text to add more information to the other sources.

The various sections dealing with the various media types i.e. sections 5.3.1-5.3.5 have all included examples of the different uses of information in business. These are many and varied and are often the means to promote goods and services to potential customers. Also the internal function of a business relies heavily on communication between the various members of the workforce. The examples have largely been to illustrate specific use of the individual media types outlined, but it is clear that often the use of more than one type of information medium is common. Frequently all, or nearly all, the different types are used together, for example in presentations where image, video, sound and text are often combined to convey

Table 5.3 Business Communications and medium used

Business Communication	Primary medium	Secondary medium	Other
Reports	Text	Image	
Work instructions	Audio	Text	Video
Training material	Audio	Text	Video
Catalogues	Image	Text	
Sales presentations	Audio	Image	Text

different information about products or a company. What should be clear is that the use of all information in whatever form should be appropriate to its purpose.

Some typical examples of business information and the types of medium used are given in Table 5.3. These also show the primary medium used and other secondary uses of other information types. This reflects much of what goes on in industry although there are some particular companies that have policies that allow specific differences to this generalisation. For example, some companies make extensive use of training video and do not have in-company trainers in addition to these resources. This is particularly useful for companies that have a repetitive training need, such as service companies where products may be manufactured by other organisations.

In terms of the technology used in business there is little evidence of a great rush to use multimedia communications technology except in a few cases of specialised companies. The majority of business and industry is still to be convinced of the benefits of new technology and its installation in a large number of companies will take some time.

There are some interesting cases of technology being used by many companies. The rise in the use of fax was such a phenomenon in the 1980s. This enabled users to send text and image documents over a normal telephone line and reduced the time between sending a document to seconds rather than days. This development is now firmly embedded in business culture so much so that the fax machine is now viewed as an infrastructure product in the same way as the telephone is. The rest of multimedia communications technology has a long way to go to reach this level of acceptance.

The next most popular communications technology in use is an Internet connection, either provided by a service provider or as some space on a general Internet server or provided through the company itself via a communications provider. The use of the Internet by business has grown rapidly in the early 1990s and is predicted to continue.

Finally, the use of higher speed connections for data communication is likely to be of increasing concern to many businesses over the next few years. With increasing demands for information and better quality of communication required the connections at individual companies will need to be improved to cope with the added requirements.

5.4 BUSINESS ORGANISATION FOR THE INFORMATION SOCIETY

Having looked, in depth, at the various uses of information in business there is now some scope to develop a speculative look at what business can do in the future. This is speculative because any number of factors can change and bring about a completely different scenario. However, the possible outlook for Europe and other Western civilisations is beginning to look very similar. Some of the trends of previous decades and centuries are pointing towards a society that is mainly based around the use and trade of information, the "Information Society". While this is often conceived as a single concept, it can hold many different forms that are substantially different, but this book is not intended to be about society so these details are mainly left out. However, the shape and culture of society will largely determine the role that business plays within it.

The information society is mainly an evolution from the previous industrially-based model. The manufacture of goods and the provision of basic needs has become increasingly automated and fewer people are required to work in basic industries to provide for the needs of the population. There are, however, increasing needs of an informational nature. The society in which people now live, at least in the western world, has a huge requirement for information in all forms and this fuels a vast industry that is in place to provide it, process it and reshape it into consumable "product".

On the other hand, business has grown up using the models of the past, the model of mass production. This was a valid model for the industrial phase of society but increasingly looks out of place in an information society where individuals are often responsible for processing information and more individual approaches are made possible by the very nature of the information technology used. There are many factors affecting the way society will develop but the general trend of more use of information is one of the most likely characteristics of any future society. To react to this new paradigm business will need to change. Old mass production practices are no longer appropriate in all cases. The business of the information age will need to use the information and communications technology to interact with customers, suppliers and its own workforce all of whom could be in other locations. This is often referred to as the virtual enterprise. It is made reality with the aid of multimedia communications technology.

Another factor that needs further consideration will be the way people work. At present the main model of work is based on the industrial model of moving the workforce to the place of work. This is, again, becoming less necessary as communications technology improves to the state where it can be used to provide as good a communications channel as a face-to-face meeting. This then leads to other questions about the relationship between the home and the workplace which will need to be answered for the technology to be successful. This is discussed further in section 5.4.1.

5.4.1 The networked world and "teleworking"

The advent of the information superhighways has been heralded as the way to reduce the travelling between work and home that has been a characteristic of the modern industrial age. This ability to work from home using telecommunications links to contact other members of the workforce and other companies has been called teleworking. It is a concept that is born out of the information age. It is true that some workers have been working at home in the industrial society but this is mainly a minority of the workforce for any particular industry. Teleworking is based around the information worker and when the main work carried out is the processing of information it is clear that much of it is place-independent, so it can be accomplished anywhere as long as the worker has access to the relevant information. The home is seen as a convenient place to work. Teleworking is the ability to work at a distance from the office using telecommunications as the medium of communication between the worker and the office or workplace.

Teleworking will involve different technology for different jobs. A computer programmer may be able to telework with a simple modem connection unless the volume of data needed is excessive which would probably require ISDN connections. If the company required contact with the teleworker then ISDN could facilitate video telephony and application sharing. Other staff may only need access to small amounts of information each day and simple modem connections are often all that is required to telework. However, if the amount of teleworkers in the company grows then the need to video conference will also be likely to increase requiring more capacity in the connections to the teleworkers. What is, therefore, clear is that any introduction of teleworking would need careful planning to see that the correct systems were installed at the beginning and not equipment that would require upgrade after a short time or a small expansion of the teleworkforce.

In addition to the communications equipment consideration is the question of how the information is organised at the workplace. Many companies use LAN segments based in departments with servers connected to a backbone. This allows all the on-site workers to access the main servers but often the individual segments are blocked from each other for security and privacy reasons. To connect teleworkers into the corporate LAN would require that they can access the information they require to do their work and that they do not have access to other sensitive information if they are not required to access it. They also need to have access to the same software that they would use at the workplace and there are, therefore, licensing issues to be resolved with this approach to work. It is difficult to give definitive statements for any teleworking situations because all organisations will be different and the use of teleworking will be unique to the company. The technology is available to do nearly any task that is required and many jobs can now be converted to telework.

Probably, even more fundamental than the technological issues outlined above are the social issues involved in teleworking. There are many reasons for wanting to switch to telework, both for the benefit of the individual and the company. The switch can have both positive and negative effects. Some of these are listed in Table 5.4.

The ability to successfully telework depends on many factors, some of these are technical and relate to the job that is being done, but most of them are factors which relate to the person's interaction with others and the workplace. Many workers feel that a workplace is a necessary condition of work and use home for different purposes that are mutually exclusive. In cases such as these it will be difficult to persuade the people involved that teleworking is beneficial. Often the proponents of teleworking only consider the benefits in terms of job efficiency and ignore these personal factors.

Table 5.4 Aspects of teleworking

Positive aspects	Negative aspects
Higher efficiency	Less face-to-face contact
Less commuting	Social isolation
Reduced work stress	Increased home stress
Less expenditure on transport	More expenditure on heating etc.
More time freedom	No clear divide between home and work
Able to work anywhere	Need to provide space to work
Able to tune work environment to suit	Difficult to tune in to corporate ethos

What is evident in much of business is that there is a good deal of resistance to change and that the old models of work will take some time to be replaced by any new practices. This is partly because it needs a cultural change on behalf of the business and its workers to change from a traditional model to a working pattern that includes telework. More discussion of the issues of telework and mobile telework can be found in Dix and Beale (1996).

Apart from the immediate social and economic considerations of telework are the long-term implications of a change to the structure and culture of work. This has been discussed previously in a number of books among which is *The Third Wave* by Alvin Toffler. This predicts that the home unit will become the focus of industrial production and consumption and the individual will be a "prosumer" devolving control from the industrial scale down to the scale of the individual settlement or home. This is seen as a direct consequence of the growth in teleworking, whereby more workers are removed from the need to commute to work and perform their task in the home environment. These aspects of telework will be considered again in Chapter 8 when a more detailed analysis of the home environment is outlined.

5.4.2 Business use of the Internet

The Internet was nearly exclusively a government, education and military set of networks until the early 1990s. Then the invention of the hypertext transfer protocol (http) and the World Wide Web increased interest in the Internet and particularly in its ability to be used by business as a medium for information transfer to customers i.e. as a form of advertising and product update. This was earlier a difficult problem since previous protocols such as email and file transfer required that the user was actively involved and needed to request information. There were also no suitable facilities for making simple information available to Internet users with simple text-based interfaces. The use of http allows users to browse information and retrieve items of interest, previously this required file transfer and browsing was limited to file directories with all the information in filenames and indexes which required separate downloading.

One means of providing information via telecommunications links was the Bulletin Board Service (BBS). This has been used by a number of manufacturers in the past to provide product update and other information along with program download and other facilities. The usefulness was, however, limited by the cost of the call involved. Frequently, the BBS would be a single computer linked to a telephone via a modem where users could dial in and scan useful information or leave messages for later answer. The advent of http has allowed this service to be offered via the Internet with direct access to the information being much easier through the commonly available web browser interfaces.

The use of the Internet has also spread much further than the BBS systems having attracted more diverse sources of information than providers of BBS systems. The simplicity of the interface and the common access protocol must have a large part to play in this. The Internet now has a wide range of companies offering services that would never have been likely under a BBS. The range is from a wider field than the obvious computing-oriented companies that were the first information providers. There are now many companies providing access to goods and services of a highly diverse nature. Internet shopping is now a real possibility with access to advanced product information in multimedia format. However, there is still little use of the Internet for company-to-company information transfer. This is partly a lack of faith in the security of the Internet for sensitive information and partly a lack of users, since there are many other company-to-company information service providers that have been established for longer than the Internet has been considered as a tool in business. These providers offer a service that is tailored to business needs and this cannot yet be said of the Internet and its many service providers.

There is, perhaps, a more crucial reason for the lack of commitment to using the Internet for all business needs and that is one of "culture clash". The Internet has grown in a fairly haphazard way from its beginnings in the research and education area. This has led to an Internet culture based around protocols that are formulated to meet immediate user needs and then adjusted later. While this provides a means of communication that is quick and effective it does not always provide a suitable means of secure and foolproof communication that is required by business users. The business culture of providing a service to customers and using appropriate tools to provide the service that are of a stated quality is a direct opposite of the Internet approach. There are, however, moves in the Internet community to tighten up its approach to service that would make it more appropriate to the needs of business.

The use of the Internet by business users is also a cause for concern amongst many business users. The Internet allows wide-ranging information browsing much of which is unrelated to business use. It is therefore not surprising that Internet use is seen as a problem by many managers in industry. The problems cited include:

- Staff time wasting
- Security of communications
- Staff shopping on-line
- Management of the user links
- Possible criminal activities by staff

Although it may be possible to argue that a responsible workforce would not use the Internet for this type of activity it is always possible that a user would use an employer's access to the Internet to disguise some of the more dubious activities that occur on networks.

The development of information superhighways will have an effect on the business use of the Internet and other networks. Firstly it will make it possible to access more information and better quality of information, or more appropriate information. For example, a user with a higher speed data connection would be able to access on-line video for trailers to movies whereas the low speed user would be required to download and use the film clips off-line. This will allow more information to be made available to users, and because they have better access to it, they will be able to use it more directly. This will also apply to inter-company information transfers where the users will be able to exchange information more rapidly. This will affect the process of business itself. The companies will be able to exchange product data as soon as it becomes available and modifications to products can be cascaded down the supply chain immediately. This will in turn affect the working practices and business culture that is prevalent in any industry.

5.4.3 The global economy

One effect of the increased emphasis on information as the commodity of interest to the world is the trend towards globalisation of the world economic system. The use of information as the raw material for processing and the ease with which it can be moved over the world in a few milliseconds has led to much information processing being moved to lower cost economies, such as Third World countries and the former communist bloc.

The use of computer networks and the mediation of communication by computers allows anyone in many countries to process information from any source, especially with low cost communications and processing. This then allows large companies to move operations around the world at will, finding the source of the lowest cost processing, often by using bid tendering as the process of distribution. This has the effect of increasing the globalisation of the companies involved and providing work for countries that have traditionally been at the lower end of the technological scale. It does, however, have an impact on the lives of employees in more advanced economies where wage rates and high standards of living militate against them.

Again this may be seen as evidence of the new organisation of the world for the information age which allows work from home or anywhere in the world giving impetus to the notion that there will be a new era of home production as mentioned in the last section. The major factor in all these predictions and prophesies is that information will be the commodity used by most people to earn a living.

5.4.4 Information and production

The main prerequisite for any large scale change in work and living practices will be the infrastructure that is in place for accessing the information and moving it about the world. The current infrastructure based around the Internet and other networks is not of sufficient capacity to justify any large changes in this direction. However, the infrastructure is being slowly installed that will enable much more interaction between remote users and information services.

The ability to produce or process information will depend on how the product can be distributed to its intended purchasers. The speed and capacity of the networks for access to information will be crucial in determining the saleability of information product and until the large scale installation of the information superhighways, information production and processing on a large scale will not be feasible. It will, however, be possible on a smaller scale and for lower volumes of information, as envisaged in many of the teleworking cases previously discussed.

For large numbers of people to become information producers and function in a tailored environment the network infrastructure needs to be pervasive and of sufficient capacity to enable the volumes of information to move quickly. This will include all the studied media types and combinations of them in multimedia communication formats. This is a prerequisite to any large change and will only happen with directed investment in the communications infrastructure at the right time. Much of this is happening in various countries, but often in a haphazard fashion that does not lead to an overall improvement in the total network provision. There are also commercial interests that require different outcomes from the installation of networks and do not necessarily install capacity for future information needs of individual consumers or producers.

To change the working practices of a large portion of the population will take some time and the provision of the infrastructure is a big part of the problem, but only a part. There are other factors which, although individually minor will add up to much larger barriers to change in the future. For example, the cultural change required for adherence to a new work system is quite profound and many individual users will be unwilling to change to accommodate this new type of working, being adherents to the familiar system of organisation that is currently prevalent.

The common features of work in the new and old paradigms is quite simple. All work is a form of "value-adding" to some raw material. The industrial society makes objects from raw materials and uses mass production techniques to simplify the process of adding the value. The information society will use information itself as the raw material and add value to make the information resaleable in its new form. So, the information worker will need access to high-quality information at an appropriate rate to enable a sufficient level of value-adding before onward sale.

5.4.5 Issues for the future

The installation of communication networks poses a few problems. These are not technical problems, mostly these have been solved or are capable of solution with technological changes, but the social problems that have been hinted at in the preceding sections and will be returned to later in the book. For the network culture to become the preferred way of living a major change will have to take place in all societies, this may or may not go smoothly, it is impossible to predict.

The access to more information is always seen as an improvement and often it is, but there are situations where the nature of the information provided has a negative effect. This is particularly problematic where there is no regulation of the information content of networks and associated services. A particular example of this is the widely reported use of the Internet for pornography and political extremism. If the networks increase in capacity and the information is more widely available then these types of problem will only increase.

What may be a more profound problem is the way society itself is organised. Currently much of society is based around work and models of production provided by the industrial society. This may not be an appropriate model for the information society which is envisaged with the provision of information superhighways. Indeed, it is difficult to predict what would be an appropriate model since we do not yet know how we would want to live and be organised in such a system! What is clear is that any change is likely to be gradual with the introduction of new models of living to accompany the introduction of the technology. Some will work and some will probably fail. There is no guarantee of success.

If the emphasis changes from a work environment to a home-based production-consumption system. It will be difficult to maintain the current system of remuneration, linked to individual effort. The basis of work in the home will be a matter of individual arrangements with the occupants and the effort of individuals will be part of the whole home production. More of these features will be discussed in Chapter 8 concerned with the home.

Although a uniform picture has been mainly portrayed so far this is a long way from reality. There are currently many areas of the world where an information highway is far from everyday needs. Basic necessities are still difficult to obtain and all effort is required to sustain life. These are the places referred to as information have-nots. These will not be able to invest in information infrastructure. Their needs are more immediate and are likely to remain so while parts of the rest of the world change yet again. The problems caused by this and other disparities are similar to those of the industrial age where many countries and regions were not able to benefit from the introduction of mass production processes.

Finally, there may be some lessons learnt from the industrial age that will help with the introduction of the information society. The problems caused by the introduction of industrial practices causing widespread disruption and deprivation to many should not need to occur. The use of information as a resource enables much wider participation in the production process and the installation of multimedia communication networks will aid the distribution further. At this stage, however, it is over-optimistic to assume that no problems will occur.

FURTHER READING

Many of the books on this subject either cover technical aspects of business communication or look at the social considerations. The first reference is in the second category. The other reference is more technical.

A. Toffler, *The Third Wave*, Collins, 1980, ISBN 0-00-211847-5.

A. Dix and R. Beale (Eds.), "Remote Co-operation: CSCW Issues for Mobile and Teleworkers", *Lecture notes in Computer Science*, Springer, 1996.

EXERCISES

1. Chose a business with which you are familiar and list all the different forms of communication that are used

    a) inside the business
    b) between the business and other companies and organisations.

    Evaluate the possible use of electronic communication for each of the items listed.

2. Consider a process such as the purchase of an item of clothing or something else of interest. Calculate what information is used between the various suppliers and customers in the supply chain from the raw material to the end user.

3. Consider a job, such as lecturer, stock clerk, personnel officer or any other and consider what might be required to enable that job to be performed by a teleworker.

4. What issues are relevant for evaluation when a new local cable network is installed in a neighbourhood. Try to list any technical and social considerations and then evaluate how these affect the installation and the operation of the network.

6

MULTIMEDIA COMMUNICATION IN EDUCATION

Summary: The use of multimedia communication in the learning environment. The use of different media types, network and communication requirements. Application examples including "The networked academy" and teleteaching.

6.1 EDUCATION, COMMUNICATION AND TECHNOLOGY

Education is a fairly obvious example of the use of communication to achieve a goal. The whole of education depends on effective communication between teachers, students and others to facilitate learning. It is, therefore, only natural that those involved in education will look towards new communication technologies to enhance the process.

As will be seen in the following sections there are many different situations in education that are used to suit the teacher and the learner. Most of these can find some advantage in the use of multimedia communication technology. However, as in the case of business communication there is a need for a comprehensive analysis of the particular situation before applying technological solutions. It is always possible that a simpler approach can be more effective and in many cases more appropriate. What is important is that the educational process is successful. This can only be so when the learner actually learns. To start any analysis the various factors that affect the learning process must be considered to some degree. Whereas the following is not intended to be a complete list it covers most of the fundamental areas that need to be considered.

Some of the factors affecting the learning process are as follows:

- *The item being learned (skill or knowledge).* Some things are easier to learn than others. The complexity of the learning and its relevance to the learner will have a bearing on how the learning proceeds.
- *Prior knowledge, attitude and aptitude of the teacher and learner.* The different educational backgrounds of learners have a bearing on the effectiveness of a learning process. For example, readers of this book who have undertaken a separate course in the technology of computer communication will be able to apply that knowledge when learning from this book. For others, the process of learning about the subject will need to encompass a wider range of material.
- *Previous learning experiences of the participants (together and individually).* In most educational situations the learner will have some prior experience of learning something similar. If that experience has been a reasonably fruitful exercise then this may enhance the later experience. Similarly, a negative learning experience can be very damaging. Also, the interaction between the learner and the teacher will play a part in the effectiveness of the process.
- *Time and place of learning.* Most readers will probably have experienced the effect of time and place on learning. Learning is often less effective if the learner would prefer to be elsewhere. This is often the case on a Friday afternoon!
- *Expectancy of the participants.* There can be different levels of expectation between learners and teachers. This can adversely affect the learning process. It is possible to ameliorate this affect by effective prior communication of the objectives of the learning process but there is still room for disparity between the expectations of the teacher and the learners. It can also happen in both directions, i.e. where the teacher expects more to be learnt than the pupils are capable of, or where the pupils expect more from the teacher!
- *Physical condition of participants.* The physical comfort of the learner is often taken for granted, but it remains one of the fundamental necessities for learning. Without a reasonable degree of well-being it is difficult to learn effectively.
- *Technological aptitude for any technology being used to learn.* Finally, and probably of most relevance to this book is the attitude towards, and the aptitude for, the use of technology by both teachers and learners. Many projects use new technology in education when the teacher or learner do not have sufficient experience in using similar approaches to the one proposed. Adapting a teaching or learning approach to a new technology can often be a useful and rewarding experience but in some situations it can actually hamper the learning process. If the learner or the teacher do not have the aptitude to use the technology appropriately the process may be less effective than previous methods of achieving the same results.

It is, therefore, essential to consider these points when proposing any use of multimedia communication or other technologies in education. There will also be other factors to consider that will affect the application of any technology to the process. An investment in a new approach to teaching or learning needs careful consideration to justify its use. Many of these factors will be connected to the technology and are similar to those affecting any application of technology. For example, the following list is a fairly general set of factors affecting the application of technology.

- *Cost.* It is relatively easy to propose a technological solution to an educational problem, but if the solution is too expensive it has little chance of adoption by the proposed users.
- *Effectiveness.* Any technological solution needs to be at least as effective as any previous method of learning.
- *Appropriateness.* The technology proposed needs to be appropriate to the needs of the learner, the setting in which they learn and the overall situation that forms the context of the learning.
- *Ease of use.* The technology will need to be as easy to use as is possible, especially when it is considered that the alternative traditional approaches of using text and voice communication are a relatively normal part of life for most learners.
- *Restrictions on applicability.* Any restrictions on the applicability of the technological solution should not preclude any necessary aspects of the learning that could be expected.
- *Restrictions on access.* Use of technology should not restrict further the access of individuals to the learning material. In many cases it can have the opposite effect i.e. opening up access to more individuals than a traditional approach.
- *Control of the learning process.* The necessary control of the learning environment should be available to any interested parties i.e. teachers and learners, where appropriate.

These will apply in many cases where multimedia communication technology is proposed for educational use and its use will need to be justified in terms of many, if not all, of these factors.

With both the educational factors and the technological factors to consider there are many constraints on the application of technology to the educational process. It is, however, a major area of use for multimedia communication technology and is likely to remain so. Why this is so should become clear on reading the rest of this chapter.

6.1.1 Range of educational settings

When considering the use of multimedia communication in education it is easy to consider a restricted subset of the whole of the educational experience. This is unhelpful as there are often many applications of communications technology that are of use to learners in different circumstances from a standard norm. The totality of different learning requirements is immense, some of these can be met with the application of multimedia communications technology, some are better served by other means. To make any judgement the range of educational needs and settings needs to be explored in detail.

Most educators now consider that learning continues throughout life and that once the process is learnt the learner can become more independent and learn subjects that are of individual interest. This sets the application area at its most extensive. The educator needs to consider the learner at every stage of this learning cycle, i.e. from the earliest learning years until the end of life. The needs of learners at all of the stages in this life-long educational process vary from stage to stage and from learner to learner. There are also variations in the needs of individual learners which arise as a result of other factors such as disability or learning problems. These can often be aided by the use of technology but do not always have simple solutions.

The range of standardised educational settings limits the need to consider every learner as a particular special case. The use of schools and the classes within them, the standardisation of curricula and the subject approach to learning all tend to standardise the educational process for the majority of learners. It would be possible, but impractical, to consider all learners as individual cases. The following paragraphs will, therefore, look at the typical cases and some individual special cases. The typical cases are listed in Table 6.1 and the special cases are covered in section 6.1.2.

Table 6.1 Educational settings

Educational stage	Typical age of learners
Pre-school	0-5
First school years	5-11
Secondary education	11-16
Tertiary education	16-19
University and college education	19-21
Adult education	21+
Postgraduate education and research	21-30
Professional updating	25+
Non-vocational education	Any

The range of standard educational settings given in Table 6.1 is typical of what happens in the UK. There are differences in other countries but the general pattern is similar. In general the stages are the same but may happen at different times and may not depend upon age as strictly as appears from the table. There is also a different emphasis in each country upon the different types and proportion of education that are provided and the number of places available in each. This does not affect the needs of the learner but may affect the effectiveness of the learning process. For example in some countries the normal size for a university course might be 100-300 whereas in others it could be an order of magnitude higher with up to 3000 students studying a particular course. These differences, while making it difficult to generalise do not substantially alter the analysis. The following sections look at each of the identified stages of the standard educational process and outline the various needs of learners in these settings. This generalisation gives some idea of the needs of learners across the whole educational range while leaving individual differences and requirements out of the analysis.

Pre-school. Approaches to pre-school learning vary widely according to the ability of parents and others that care for young children, to provide for their learning needs. The general requirement on most young children is for social skills at this stage of the learning process. These will then assist in the later stages where there is a more formal requirement to learn knowledge and skills. Most learning at this stage is highly individual and depends on close contact between the parent and child. It is unlikely that the application of any technology will enhance this process although, in some situations, it may aid in the communication between parents and children who cannot be together for some reason. It is highly doubtful whether this would be a useful sole means of educating at this stage.

First school years. The first years at school are generally used for basic skill learning to facilitate further learning. That is, this stage of educational development is used to acquire a basic skill level in reading, writing and communicating that will allow the learner to develop in other areas and subjects. Without this basic grounding in these skills the rest of the stages are largely superfluous. At later stages of primary education the boundaries between subjects start to become outlined in preparation for a more subject-based approach in later stages.

Secondary education. The secondary stage has previously been seen as the end point of education for young people. With the advent of life-long education this has become less so. There are, however, some that leave formal education at this stage and do not return. This requires that some degree of attainment be noted so that their standard of education is recognised by employers or other interested parties. For those that continue, the need for recognition of achievement is determined by their proposed course of action in the next stages of the educational process.

Tertiary education. A tertiary stage of education for young people is often determined by the learner's ability and aptitude. This can also decide their subsequent path and future needs in terms of education and learning. This is not to say that this is the only course of action but to drastically change course at this stage would require a probable repeat of this stage in the process.

University and college education. Similarly to the tertiary stage the higher education establishments have an increasing tendency to specialise in subject areas and this is largely determined by the learners' ability and aptitude for particular subjects. There are some exceptions to this where a broad range of subjects can be studied at this level, but most students tend to chose those that have been studied before or those that lead to specific career paths such as law or medicine.

Adult education. After the more common stages of education (i.e. those outlined above) there is more diversity to meet the needs of individuals. Some learners need even more specialised knowledge, some need to revisit earlier learning experiences to recover skills lost or never properly mastered. These latter facilities are generically available under the banner of adult education, where learners can gain skills from the most basic reading, writing and communication skills to more advanced levels of learning.

Postgraduate education and research. After a university education the options for further study are split between further taught provision such as a master's degree or the option of engaging in research. There is an element of specific research training i.e. training in the process of research, but the predominant theme is the individual thesis on a topic that is to some extent novel. This requires highly specialised input and resources.

Professional updating. After a number of years of employment it is often found that professional workers may need some theoretical input to update skills, especially in highly complex industries where standards and techniques can change quite rapidly. This is common in many industries and allows a broadening of subject knowledge that is often not traditionally possible in the workplace, or only to a limited extent. Professional updating allows access to a wider range of expert skills and knowledge than is usually available in a workplace.

Non-vocational education. One of the largest areas of education is the non-vocational area where the learner is studying for personal interest and enjoyment. This is most often done in the learner's spare time and is seen more as a leisure activity than an educational one. It does, however, have the same characteristics as the other forms of education even though it may have a more dedicated or interested set of participants.

All these different educational settings require different informational input and different levels of information to enable the learners to benefit from the educational experience. For example, the same subject is often taught at secondary school level as at university level but the level of detail required is much higher in a university. This variation in informational needs must be addressed by any provision of information services in an educational setting. An overabundance of information in a particular subject is almost as bad as a lack of information. There needs to be an appropriate provision for the learner's needs, or some way of providing an appropriate level from the information provided e.g. a filtering mechanism.

In the various settings there will also be a differing individual element that is unpredictable. From the youngest age learners have their own interests and requirements. Although this may be limited in a primary school setting, at the research level all the individual learners have their own particular requirements and there is little overlap between any two. There are, therefore, distinct requirements to provide information services to education to meet the needs of different users in the different settings that are common within the educational world.

It is, therefore, fairly clear that there can be a general information service provided for educational needs but it requires some measure of tailoring for different uses. It will also need some further tailoring for individual learner's requirements as not all learners are the same, as will be seen in the next section.

6.1.2 The individual and group learner

At all stages of educational development there are different types of learner. The differences are usually more marked when there is no formal setting for the learning process such as with a home-based learner or in distance education. Even so there is still a large degree of variety in the type of learner within a single setting or learning environment. People differ in many respects as regards learning, some learn more easily than others and some find certain ways of learning more conducive. There are also different situations that are a useful categorisation of learners based on the scale of the learning group and the geographical spread of the group. This is illustrated in Figure 6.1.

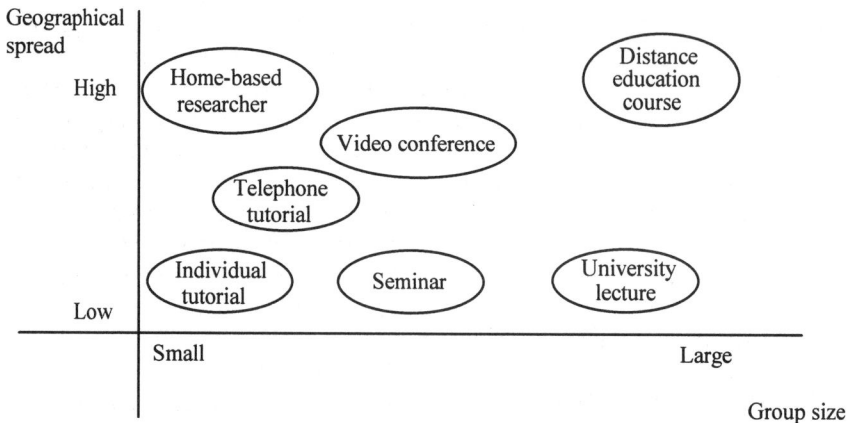

Figure 6.1 Range and scope of learners

In Figure 6.1 the range of learners from the individual to the large scale group is covered as is the spread from the relatively small-scale tutorial to the large-scale distance education group such as would be found on many open learning courses. In discussing any application of communications technology to education the whole range needs to be considered. To do this the four distinct cases indicated by the small and large group size and the low and high geographical spread will be considered. Each of these is considered in turn in the following sections with a typical example.

Low geographical spread—Small group. A typical example of this would be the university tutorial or a music tuition group. The small spread of the participants would not normally make the use of multimedia communication technology likely. However, in certain circumstances a small group of learners may want to use technology to form a group rather than arrange a physical meeting. This may be the case when they are, for example, physically disabled or visually impaired when the aid of technology assists more than in cases where there are fewer barriers to learning.

Low geographical spread—Large group. In cases where the learning group is larger the use of multimedia communications technology would appear to be less likely. However, in large groups it is often difficult to communicate effectively with individuals in formal situations and the back-up of some technology is often useful. For example, the use of electronic mail to send messages to large lecture groups is common in universities.

High geographical spread—Small group. The spread of learners has much more influence on the use of communications technology, although there may be some cost implications for widely dispersed groups of learners. This aside, there are distinct advantages to be gained by using multimedia communications technology for communication between members of these groups. The technology enables meetings and sharing of ideas which would only have been possible by travelling to a meeting in previous times. This is true of any size of group. However, the small group has the advantage that the range of technology for interaction is wider since the allowable number of participants will be determined by the capacity of the communications links. Fewer participants use less capacity. A small research group could function in this way or a student self-help group. Alternatively, a small group of dispersed professionals could use a networked service for updating. This would make the provision of such a service more effective as there would be less need for them to travel to a particular point for access to the service.

High geographical spread—Large group. A large group over a widespread area has additional problems in that effective interaction will depend on the technology available. Much large group education is done by TV or radio broadcast and this provides a cost-effective way of reaching the larger group. However, for more interaction some two-way process is needed. Computer conferencing can achieve this at present, where any number of participants can use text-based discussion groups to learn about different topics. Many of the newer multimedia communication technologies are not appropriate for use in these widespread groups. Electronic mail is an obvious exception as it can be used, as in the low geographical spread case, to contact individual or whole groups on this scale.

Individual learners will have their own needs depending on a number of factors only one of which is the situation in which they are learning. The subject, the surroundings and many other factors will play a part in the learning process. There are also many different needs that may be unique to individual learners. These factors will also play a big part in the success of the process. Although there are too many individual needs to outline here it is worth noting some cases that are common.

Individual special learning needs can arise in many ways. Disability, physical isolation or remoteness, special requirements etc. What is common is that there is a difference between what is normally considered necessary and the individual's learning needs. For example, a partially sighted person may need a large-print version of any text used in the learning process. Often these needs can be satisfied by using multimedia communications and the differences between any standard provision and the special requirements of individual learners can be minimised often by variation of the software used to communicate. For example, it is possible to use a speech synthesiser to convert text to speech for the visually impaired.

Finally, the way that learners use information is also important. In relatively standard situations such as in mainstream schools or universities at undergraduate level many of the learning requirements can be provided by standard texts and other simple facilities. When the problems of learning become more complex the need for access to more sources of information begins to play a part in the process. For example, the conduct of research requires access to many different sources of information and this is mostly not available in a single library. The solution to this in the past has been to loan books and other information sources from one library to another, an inter-library loan facility. This will become less necessary as multimedia communication networks and the types of library discussed in Chapter 7 become more widely available.

This type of information provision will enable individual learners to access material from many different sources using the mechanism of the multimedia communication network. This will be explored further in the next section.

6.2 EDUCATIONAL COMMUNICATION

Because of the many different types of learning environment and the multitude of different learning patterns employed by individuals there is a wide variation in the possible provision of information for educational purposes. There are also many different views on the need for different scales of provision that need not be of concern. The traditional means of providing information for education has been with text in books (like this one) and through images and more recently radio and television broadcasts. This goes some way to being a multimedia communication method of using information. However, there are now many technological possibilities that allow much more variation in the provision of information for educational purposes. This variation can be summarised by considering the extremes of the multimedia provision of information.

The provision of educational information to learners in multimedia formats is often based around the CD-ROM as this was one of the first effective means of providing the different types of information in a digital form for use in a computer. This concentrates all the information in a single source, the CD-ROM, which can then be used to provide educational input. The other extreme is the provision of information via computer networks where individual information sources are more specialised and the software to control access to these resources is based mainly on the user's machine. These two extremes have been called the *media-complex model* and the *communication-complex model*. Fuller details are in the paper by the author referenced at the end of this chapter. Table 6.2 gives a comparison of the two schemes.

Table 6.2 Comparison of media-complex and communication-complex multimedia

Media-complex	Communication-complex
Only one source needed	Many addresses required
Limited by media size	Nearly unlimited resources
Fast access	Speed depends on network
Cost related to media	Cost related to network access
Easily standardised	Easily customised
CD-ROM based	Communications-based

It is clear that the two models both have a place in the provision of educational material to learners. It is also clear that there needs to be careful consideration of the needs of the learner before deciding how to use multimedia communication to provide educational material. For example, the complexity of network provision makes it less useful in standardised classroom situations than CD-ROM, whereas for individuals preparing projects, or doing research, the CD-ROM would soon need supplementary material.

It is, therefore, likely that the provision of information for educational purposes will take any number of forms both at, and between these extremes and that those that determine the learning will be able to choose between media-based and communications-based provision. The choice will be fairly crucial affecting cost, suitability and applicability of the results. This does not imply, however, that in standard situations such as in basic education classes, that network access should not be used, but that it should be used appropriately to provide necessary information. This will inevitably mean that the director of the learning experience, be it a learner or a teacher, will have greater responsibility to identify useful information.

6.2.1 Learning and communication

So far, in this analysis, the complexity of the learning process has not been discussed. It has been assumed that communication of information is the main part of the educational process. While this may be seen to have a part in learning it is not the total picture. Many parts of the educational process are not directly communicated to learners. However, there is still a great need for information, but it is important that the correct information is available to the learner at each stage of the learning process. Again, the appropriateness of the information provision is the important factor.

Since this book is concerned with multimedia communication and the provision of information it will not consider the other factors that determine how the educational process works, but only consider that part of it directly associated with the communication of information. It can, however, have an effect on the possible approach that is taken in a particular learning situation. For example, in a school classroom the learning about other parts of the country can be a rather academic exercise without the resources to visit the area in question. With the use of a multimedia communication network the class could interact with other pupils of the same age living in the area being studied and learn about it through the aid of others (with the benefit of it being a two-way process to benefit both schools).

This type of two-way process is very difficult to organise effectively without the aid of a multimedia network and would not have the same immediate impact since it would probably rely on postal exchange of still pictures and text (possibly video-cassettes). An on-line version can be much more interactive. This type of interaction has been commonly used to test the usefulness of communication networks in schools and a number of different case studies can be found in the conference proceedings listed under Further Reading at the end of the chapter. What is clear in all the different case studies is that the benefit of real interaction, especially with peer groups, is very helpful in learning situations. There may, however, be an element of uniqueness in the exchanges that makes them more useful as an experiment but that may only become apparent when the networks are more routinely established than at present.

6.2.2 Educational information—use of different media

In Chapter 3 the relative complexity of information was analysed to show how there is a structure incorporated within it to allow different levels of information to be extracted from raw data. This is especially important in educational settings where appropriateness of information is particularly important. There is no value in overloading learners with too much information at an early stage it can confuse more than it informs. It is, therefore, useful to be able to present information in an appropriate form for the needs of the learner by using this structure and the media that are most appropriate to communicate it.

It is common that the different educational settings mentioned in section 6.1.1 use different approaches to teaching and learning that have been derived to meet the needs of the learners in that particular environment. This already to a certain extent takes account of the inherent structure of information in using the sources available to best effect. The introduction of multimedia communications technology will enhance this choice and give more choice in the available media and information forms available, making the problem both simpler (because there is better and more appropriate information available) and more difficult (because there is more choice at different levels of information content).

So each of the stages of learning, whether in standard educational or not, can have access to appropriate information using multimedia networks. This information needs careful filtering to allow access at the appropriate level to different groups or individuals at different learning stages. Many current systems only provide information in a broad sense without any "value-added" service to provide tailored information. It is usually left to client-end systems or the user to filter out unnecessary information. Whereas, this may be an appropriate method of information provision for the independent researcher it does not enhance the prospects for young school children or their teachers as the necessary work will only add to already stretched resources (i.e. the teacher's time).

6.3 EXAMPLES OF EDUCATIONAL USE OF MULTIMEDIA

The different possibilities in education are concerned with the particular audience in a learning situation and the interaction between the teacher and learner, although there may be occasions when there is no teacher present. The different possibilities have already been shown to consist of a spectrum of different mixes of media and local or remote access to information. This does lead to some difficulties when the whole range of options is considered. For example, the use of video is much easier in a local environment on tape or CD-ROM where there are no significant bandwidth restrictions on its use.

In other more specialised situations there may be a necessity to access a more remotely located item where it is the only useful item available. Again video provides a good example. If a researcher needs to observe wildlife in a remote area a video record can provide useful information, it may be possible to use cameras installed and watch at a distance. (In 1995 an interested Cu-SeeMe user concealed a camera near a bird's nest at his home in the USA and it was possible to watch the nestlings grow from anywhere on the Internet.) Generally, quality video requires too much capacity to be used in real-time at present and the quality or the timeliness suffers.

The media-complex and communication-complex models will both be relevant in future uses of multimedia in education. What is important is to find the most appropriate model for each individual learner or group of learners and to enable that to be used. As any reader will realise this is not always possible so some compromises will inevitably have to be made. Each learning situation will require some adaptation of standard provision, with a few being able to make full use of simple, standard offerings.

A CD-ROM will provide a large amount of information in a compact form but there will inevitably be some extra information required. This can be provided in a traditional form or in electronic forms via networks. The use of CD-ROM will be applicable to many standardised situations such as standard curricula and general information provision. Some CD-ROM will be applicable to more specialised areas where the information is definitive and can be easily handled using the CD-ROM format.

Much information is not as simply dealt with. The possibility of change and the requirement for currency of the information requires that a more alterable technology is used and the best way of providing this type of information, at present, is to make it available on a network for browsing or download by users.

Both CD-ROM and networked information have the benefit of interactive access so that users can peruse information and navigate through it with the appropriate software tools. At present the two schemes are most likely to be entirely separate, however, information provision is likely to be enhanced by combining the two distinct approaches into a hybrid approach to suit the user or learner in any particular environment.

6.3.1 Network availability in education

To be able to use appropriate information for educational purposes requires that the information be available. If a CD-ROM of the information is published then that is usually a good source. If there is no CD-ROM and the subject is relatively new then the use of a network will probably be the only likely route to successful information retrieval. This then supposes that the network will be available and the information will be accessible from the user's access site. This is not necessarily the case as some restrictions and problems can occur with networks and their operators.

In all educational settings the access to appropriate information is vital, it is no use having access to a research page on physics if the learner is only meeting the subject for the first time and requires a digest approach such as would be used in a school. It is, therefore, clear that the availability of the network and access to it are only part of the whole picture. To make information available on a network there needs to be a definite requirement, a willing person or organisation to make the information available in the required form and a suitable means of advertising its presence to the intended users and, maybe, restricting access to other groups. These four areas will be discussed in the following paragraphs.

The requirement. Identifying information requirements of learners is usually relatively straightforward. It has been an integral part of teaching for many years and the skill of many teachers is their ability to identify precisely what information is required for a particular learning situation and which can be left out. Therefore, building on these skills will be the best way for information to fulfil the educational requirements of the learners using either the networks or the CD-ROM approach to multimedia learning. In the network environment the extra skill required to collate much of the information that is currently available and present it, or pointers to it, in a useful form. (Further details of this approach will be found in Chapter 8.) The requirement cannot always be met locally. Often a consortium approach needs to be taken where a number of educational bodies collaborate to provide information of use to them all. This can then be distributed around the network as appropriate with links between the various sites to allow access to the whole of the material. This approach is particularly useful where resources are at a premium and sharing of the cost of initiating information provision can ease the burden for individual establishments.

The provider. The provider of the information can be almost any person or organisation. It is, however, necessary to be able to discriminate between information and promotional material, especially in situations where the learner may be more impressionable, such as at primary school level. For example, the provision of information about the manufacture of soft drinks could be provided by a drink manufacturer or an independent person or organisation. The information content of the material may be little different except that the drink manufacturer will probably see the opportunity to sell more product by including advertising, whereas an independent organisation will not need to do this. There are, however, situations where a commercial interest is useful. In a university, for example, a network-based information retrieval exercise can use information from many different sources and the provision of up-to-date commercial product information can be useful where the subject requires it. This is particularly so in the computing industry where new techniques and devices appear frequently. Many manufacturers now provide network-based information to advertise their products and also to provide technical information about them. Generally, the information used in education is not of a commercial nature so the provider will be a teacher or someone similar, who in the past may have written a book about a particular subject. Now the information may be provided by a multimedia delivery system, such as a network or a CD-ROM.

Advertising. In addition to the problem of advertising intruding into the learning process the problem of making a resource known to its users also needs consideration. Many local resources can be effectively advertised by mail distribution lists, as can many of the more mainstream national and international efforts. It is more difficult, however, when an information resource does not fall neatly into a particular category. The use of newsgroups and mail distribution lists is not necessarily very useful since many of the intended recipients of the information may not subscribe or read groups not directly related to subjects of interest. A more general approach is then needed where a user can eventually send many hundreds of unwanted email messages to uninterested users and provoke a predictable response. So the problem is not fully solved. Collaborative efforts are intended for mutual benefit and do not require further promotion unless the original intention was for wider circulation. Commercial information provision will have the backing of some commercial marketing experience and can often be promoted through more traditional means initially. Individual efforts are likely to be much less widely known unless a comprehensive plan of dissemination is used.

Restrictions. Finally, the restrictions that can be applied to networks can be problematic. Many individual users find that cost is the main barrier to finding more useful information and that the use of the network is shaped by the perception of the cost involved. (The actual cost and the perceived cost may well be different and when judged relative to other expenditure may be reasonably small, but users often work under such misconceptions.) Information providers may charge for access to certain information which will only allow those with the ability to pay to have access to the information. Other information may be inaccessible to users through various causes. For example, use of the Internet in Europe can be dramatically different when information is accessed from sources in the US at different times. The use of the Internet during the US working day is significantly slower than when the US is mainly asleep! Users wishing to access resources in the US soon learn to tailor their access times accordingly.

At the university level and for research tasks the use of networks is likely to be the main source of information in future years. Some of the problems and areas of interest are discussed in Chapter 7, but information used by students does not always fit into the neat categories used in libraries. Some users may only require a few bytes of data for a particular need. It is, however, still a valuable item to the user. Networks allow both small and large items to be stored and retrieved and they allow a diversity of sources to be used for more current information than is possible using conventional means or CD-ROM-based information. That is not to say that CD-ROM has no place at this advanced level of education. There are many instances of useful information being stored on CD-ROM that are used in universities. For example, the use of CD-ROM to store abstracts of published papers is well-established and provides a first point of reference for many researchers. The provision of abstracts on networks would require a search tool to query them and a network infrastructure capable of querying many sources of information.

There are some information providers that now use networks to distribute information. It is possible to have the contents pages of academic journals automatically emerald on publication. This assists in research as the papers can be subsequently ordered from the local library. There are also some fully electronic journals in some subjects.

6.3.2 The use of networks in education

The Internet may have started life as an university, military and research network (ARPANET) but it is now much more widespread than these communities and has developed into the major focus of attention in networking. On the Internet any connected computer can be available to any other (barring explicit restrictions) and this allows masses of information to be made available to Internet users relatively easily, especially with the use of http (hypertext transfer protocol) and WWW (World Wide Web) browsers (see Chapter 8). However, the Internet is only part of the picture, local area networks also play a part in the information provision in most situations. Indeed, best use of the Internet can only really be made when the access method links a local network to the world-wide sources of information on the Internet. This may not always be economically possible.

For any educational network installation there needs to be a plan for information provision and retrieval to allow best use to be made of any expensive connection time and to ensure local resources are fully utilised. For example, any resources needed often for a class of students is better provided locally to minimise access costs and times, this can be done on a shared CD-ROM or on a standard hard disc file server.

There are, however, times when information is best retrieved or exchanged with remote sites. This usually applies when the currency of the information is important or there is a need for interaction between users. This could be between teacher and learner, between different learners or between teachers. Some of these interactions are discussed below with some illustrative examples.

Communication between teachers and learners. The traditional interaction between teacher and learner has been face-to-face. There are also some common instances of distance education where the learner has interacted with the teacher via books, audio cassettes and video tapes. The installation of networks for communication between teachers and learners allows many different types of interaction. Some will inevitably be more successful than others. The networks could be used to extend classrooms by using video links to other sites, allow teachers to work from home to teach to a number of different classes in different places or allow the learners to be separated individually from the teacher giving a remote, scattered group the same opportunity to discuss and interact as a group in conventional setting. As an example the following is taken from a study undertaken in Italy (see Coiro *et al.*, 1993)

Example: Remote learning. This example is of an ISDN link used to link classrooms together. The classrooms are in different towns separated by about 45 km, although they could be at any distance. The system allows transmission of bi-directional full-motion compressed video, bi-directional audio and high resolution still images. The system replicates the traditional classroom but allows for a distributed audience to the lecture. The ability to manage still images over the link allows the lecturer to use images in addition to audio and video to illustrate the lecture. The difference between using the still image and the video is that the still image is transmitted at slower speed and gives higher resolution than can be obtained in the video mode of the transmission. The system, as tested, gave acceptable quality with relatively low capacity with ISDN 2 being the minimum acceptable level of service required. This sort of link and use of communication is likely to expand in future as more students require access to education. With equipment of this type it is also possible to record lectures for later playback, but this does lose the interactive element that was built-in to this prototype. There are many possible uses for equipment of this type and the use of ISDN ensures that the standards used will be available in different countries where ISDN is available.

Communication between learners. The traditional communication between learners has often been an informal method of learning. It is also highly valuable in some instances. Learner interaction is generally encouraged and to widen the scope of such an interaction can only benefit the individual's education. The use of multimedia communication networks will enhance and enlarge the possibilities of learner-learner interaction making it possible to speak with and see other learners in many different countries, rather than just the few other local learners, such as classmates and fellow students. The current links between students are fairly limited since much of the educational use of networks does not have a great deal of money spent on it, but there are some useful initial uses of the Internet that may point to the future. One of these is the Kidlink project (see Stefánsdóttir, 1993)

Example: Kidlink. The Kidlink project uses electronic mail to link children, aged 10 to 15, together in a global network. The main mechanism is the email distribution list that forwards mail to a number of recipients. Some of the recipients are mail exploders giving a local distribution list facility to save on Internet capacity. The different mailing lists provide a structure to the discussions with topics for introduction, making friends, sharing projects etc. Although the project started in Norway, there are now subscribers in over 40 countries and it has provided some very useful contacts between children of similar ages in very different environments. In time this type of project can be implemented on higher speed multimedia communication networks to provide better communication between participants and allow more information to flow between them.

Communication between teachers. In addition to enabling communication between disparate groups of learners the installation of networks can also enable a much wider spread of discussion amongst the educational professionals: teachers, lecturers, education-related publishers etc. This can foster a greater sense of shared community and allow a wider discussion of issues and easier resolution of problems than at a purely local level as would be the case in a non-networked environment, such as a traditional school or college. This can then be extended to incorporate multimedia communications where the sharing of information becomes more practical and widespread. For example a group of teachers would be able to discuss the implementation of different courses and curricula over a video conferencing network instead of meeting face-to-face saving hours or days of travelling time.

Example: Teachers' distribution lists and educational newsgroups. A number of good examples of collaboration between teachers and other education professionals can already be found. There are a number of distribution lists and newsgroups which can be used for educational discussions and are often related to educational topics. Many of these discuss particular issues related to specialised educational topics such as computer science or programming education; some are more general discussing educational issues; some are of interest to educationalists as topics for classroom discussion etc. There are a number of problems with this approach to sharing information. Often, the initial investigation of newsgroups and their content and the location of useful distribution lists can prove very difficult. Subsequently, the control of messages can be problematic. Some newsgroups and lists generate many messages, some of which will be interesting, but on occasions many messages may have to be read to find some useful information.

One of the main uses of networks in education will inevitably be for the browsing and download of information. Much of what is available in various ways can be considered to be of value to different levels of education. Some of the information available may not initially have been intended for educational use but can, nevertheless, but put to good use. In these terms there are a number of different sources of information that can be useful in educational situations: information from other similar educational establishments, information from national bodies with some responsibility for educational matters, commercial educational information providers (e.g. publishers) and other commercial information not directly intended as educational. The following brief paragraphs look at each of these possible sources of information.

Similar educational establishments. One source that is likely to have useful information to any educational task is that held by similar institutions, especially those that offer similar courses or adhere to the same curriculum. This could be at any level of education, but has been more likely to be at college and university level, although with the increasing installation of networks in schools the amount of information at this level is set to increase.

National bodies with some responsibility for educational matters. Most countries have a number of bodies with some responsibility for educational matters. Often a ministry of education or a government department oversees policy and other matters. There may also be independent and semi-independent bodies, such as the National Council for Educational Technology in the UK, which can provide very useful information about many aspects of education. Many of this type of organisation provide information without cost to the user to extend good practice and experience into schools and universities. There does, however, need to be a sufficient level of funding to enable them to have much impact when compared to the commercial organisations that do a similar task.

Commercial educational information providers. A number of educational publishers have expanded their media use to include information distributed via networks. Mostly this is seen as being as an additional feature of the core business function i.e. selling books. There are also some books that include additional more up-to-date information using network distribution (this book is an example!). There is, however, a substantial amount of research into how information can be provided and sold using networks as the medium. This will then enable true multimedia publishing with the network as the source of information and updated information being available continuously. At present, most commercial information providers are relatively specialised with most using expertise gained in more traditional settings to provide the background to their networked offerings. Typical examples being newspapers and academic journal publishers.

Other commercial information not directly intended as educational. The information provided by other commercial organisations can also be of use. The provision of information by any organisation is usually related to its need to trade and sell its goods or services (see Chapter 5). There is also, occasionally, value in some commercial information for educational purposes. An experiment carried out by the author revealed a surprising amount of useful information concerned with communications networks to enhance the information available from educational sources. The experiment involved the browsing and download of information about computer communication by a group of students studying on a university course in computing. The range of sources spread across both educational and commercial sources and a number of different countries of origin with both manufacturers and standards organisations being represented. Although this example is a little specialised as the networks tend to include more information about communications than most other subjects it is a pointer to how current information can be used in educational settings in future.

It will have been seen that there is no real difference between the use of networks for educational purposes than any other. Mostly the uses will be for access to information, communication between learners, teachers and others and for expanding the possibilities of education to disadvantaged groups and individuals e.g. the isolated or disabled. This can all be accomplished by using the standard features of networks that are currently available. Applications such as distribution lists, computer-mediated conferencing, video conferencing, World Wide Web browsing can all contribute to the use of communications networks in an educational setting.

What is now clear is that there needs to be some specialised effort put into how the organisation of multimedia communication networks can be used effectively for education and how the information obtained via the networks can best be used in an educational environment. Currently the networks are relatively new and their use is fairly unsubstantial. In future, it is likely that their use will increase and these questions will need some serious debate.

6.3.2 Typical applications of multimedia communication networks

To illustrate the possible use of multimedia communication networks there are a number of examples that show the current and future possibilities of the technology. These are closely related to current educational trends and are both feasible and, in some cases, already in operation on a small scale. To find out the current state of the educational applications of networks the reader should refer to the IFIP TC3 Teleteaching series of conference proceedings as quoted at the end of this chapter and the subsequent events. The examples have been chosen to give some impression of the spread of applications and include languages, geography, professional updating and the networked university.

Language learning. Learning a language has been accompanied by numerous aids from the various media types available. It is common to find audio, video and text combined in a language learning package. This can, clearly, be presented as a multimedia package using digital information, and is now available in this form where the separate media are combined into a CD-ROM for individual use. The use of networks and multimedia communication will enhance this process further with individuals being able to try speaking the new language with others in different locations. This will enable the learner to see and hear the language spoken by current native speakers in their own environment. This allows a much more current version of the language to be learnt than is possible using traditional media where the content is usually updated only at relatively long intervals. It also provides the local language teachers with a means of updating skills without frequent visits which are often difficult to schedule. A typical scheme could be as illustrated in Figure 6.2.

In Figure 6.2 the learners in the two countries are connected to local networks and the teachers are also connected to the same network. The multimedia connections would need to be managed to allow best use of the capacity between the two countries. There may also be a requirement to use specialised equipment such as video cameras and sound equipment which may only be attached to certain of the machines on the local network.

Such a scheme as this would not necessarily be used on every occasion and the connections between learning groups would need organisation so that interference in the learning process did not occur. The learners in one country would need to trade their language learning for language teaching of the other country's learners at another time. This may not be built-in to the local curriculum in either country and would need to be carefully monitored so that it did not interfere with the normal process of the courses run in either institution.

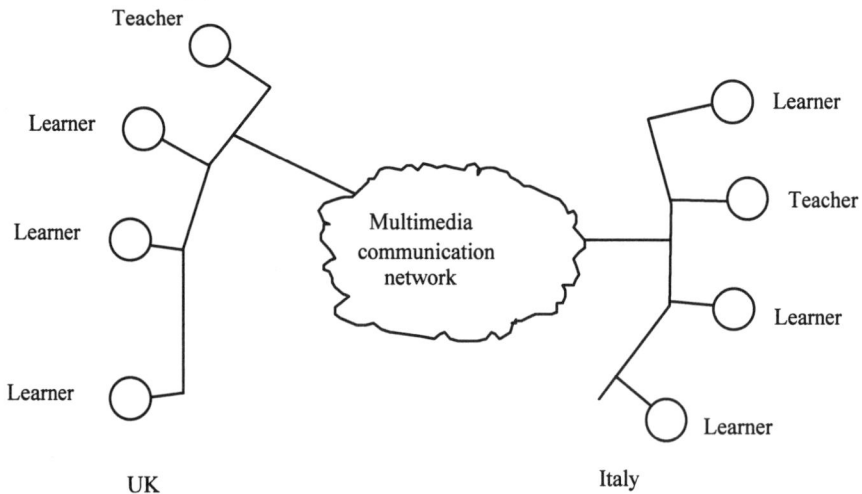

Figure 6.2 Language learning scenario

There could also be a similar arrangement for one-to-one tuition in languages where initial contact is made via email or distribution lists and then the subsequent contact is via a multimedia communication channel such as ISDN with video and audio channels. This would enable individual learners to help each other at convenient times.

One problem with such a scenario can occur if the languages are from widely separated parts of the world and the time difference makes it difficult to contact the other set of learners at a convenient time for both countries. There are usually alternatives, but not in all cases. Careful planning would ensure an appropriate partner in such an exercise.

The networked academy. The networked academy is not a new idea. It has grown out of the open learning paradigm that has been developed in the latter half of the twentieth century from the earlier correspondence courses and colleges. Partly it is a direct evolution of the ideas of distance education via text and other media and partly it is new based around networked learners, all in different locations.

There a number of possible scenarios for the implementation of a networked academy and these will largely depend on the history of the implementers. Most efforts in this field will not start from a "greenfield" site. Some will be based on existing colleges and universities, some will be based on existing distance learning centres and organisations.

The technological developments for networking are largely in place, although cost may still be a factor affecting the growth rate of such a development. As costs decrease the growth of networked academies is likely to increase, with technological developments being offered to students in the bid for growth. What is clear is that the technology is only an enabling force and much work needs to be done to support such a venture.

For successful learning in any situation a student needs a number of factors to be in place as was seen at the beginning of this chapter. For the networked academy these will include:

- Technological capability of the student
- Technological capability of the teachers
- Availability of suitable material
- Access to resources at reasonable cost
- Ability to access resources at appropriate time and place
- Ability to control the learning process.

The networked academy attempts to offer individual solutions to learning by allowing access to tutors and learning material via networks. The access offered will depend on the ability of the organisation to respond to the needs of the learner. For example, some students will be able to learn effectively with a minimum of input from a tutor, whereas others may require regular guidance to understand the principles of particular learning materials. Therefore, a range of options needs to be built in to networked learning to allow for these differences, as it is in most learning establishments at present. If the learners are remote from the tutors a video link may be the best route for access to them, with specified times of availability and other tools such as application sharing and whiteboards. Some subjects may find text sufficient, but image and video are likely to be needed in most cases.

A typical set-up could be as illustrated in Figure 6.3. This shows how some of the networked academy might function. Isolated individuals and groups of learners are able to attach to institutional networks where groups of teachers will be able to respond to problems and teach the learners. There may also be some individual teachers working either from home or at out-stations where equipment is available to establish conferences with groups of learners. There are many different possibilities, each class or group of students may well be differently organised and this will, mainly, depend on their (and their teacher's) needs and abilities.

The networked academy offers many benefits and also many problems that may only be solved when fully working installations are tried in real teaching situations. To be successful a background in distance education and an investment in technology that will enable full communication between members of the learning group should be a start. There will also need to be sufficient support mechanisms, both technologically and educationally, for any teachers and learners using these systems.

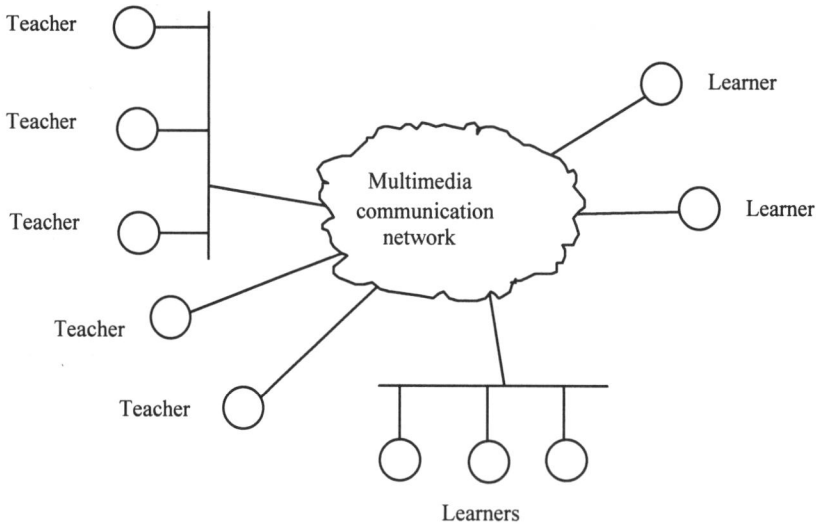

Figure 6.3 A networked academy

Teleteaching. This is a more generic term for all distance education done with the aid of computers and communication networks. A special case of this is the networked academy described in the preceding section. Teleteaching encompasses all computer-mediated contact between student and teacher using technology as a medium for communication. Examples of the scope of teleteaching have already been described: using ISDN to link classrooms, using computer conferences to enable widely separated school pupils to confer with each other, using video conferencing to consult with tutors and lecturers in a networked university.

One of the commonest examples cited for the benefit of a teleteaching approach is the distance learner. In conventional distance learning environments learners are provided with material in text form, and additionally there may be some audio and video information (probably on cassette, or broadcast for larger courses). Also, there is normally contact with an "expert" tutor. This can be via face-to-face meetings, telephone tutorials or via the postal service with written text. The important point is that there is a tutor available to answer questions i.e. to interact with the learner.

In a teleteaching environment these facilities can all be provided in a number of formats. The text, image, audio and video can be sent out on CD-ROM, downloaded from a network server or sent out in a more traditional format. The advantages and disadvantages of each are listed in Table 6.3.

Table 6.3 Advantages and disadvantages of different teleteaching formats

	Paper and tape	**CD-ROM**	**Networked**
Advantages	Easy to use Familiarity Low-cost to user	Integrated medium Common on PCs Reasonable cost	Easy to change Any format of client Can be selective
Disadvantages	Difficult to update Postal distribution Requires handling centre	Difficult to update Postal distribution Not all formats	Cost of access Speed of network access Media support

The characteristics of each of the formats in Table 6.3 should be familiar at this stage. The medium used for any teleteaching application will depend on the nature of the subject, the method of support for the learner, the access to networks and the degree of aptitude expected of the learners. The use of networks need not be assumed to be the best method of achieving the desired results; effective learning. If the support mechanism is not in place learning will be hampered and the best materials will need some support for some learners. The use of any of these methods need careful consideration weighing up the options for each before deciding on a solution. Schemes using technology for its own sake rarely succeed.

Professional update using networked information and communication. A similar scheme to the teleteaching proposal in the last section can be used for the updating of professional workers. If the information required is mainly available on-line then a continuing approach to education and learning can be taken, with the learners using the network to access material when and where required. This can enable a more continuous approach to learning than has been used in the past where professional updating has been delivered in blocks at specific locations. The use of a network will allow a more learner-tailored approach than has been possible in the past.

Another aspect of the use of networks in distance education, and especially in the area of professional updating, is that learners no longer need to consider location as of prime concern in choosing between available offerings. Networks remove the necessity to attend a specific institution or site. The network-aware user could even choose individual sections of a course, to tailor a complete package, from different providers if necessary. This assumes that the providers are willing and the necessary course validation allowed the learner to gain credit in this way. There can also be more information made available by professional bodies direct to the learner.

6.4 CURRENT AND FUTURE ISSUES FOR EDUCATION

As with all applications of technology the use of networks in education will have both positive and negative effects and outcomes. What few instances of applications there have been so far have shown that there are a number of possible uses for networks and multimedia communication in education that can both enrich and expand the learning possibilities for many users. This does not mean that they should become the only source of education. Indeed, there are situations where the use of networked teaching can be a retrograde step. In this category is the first school experience of children where the actual learnt material is only part of the wider socialisation process so important to the continuance of societal values.

There may be situations, even at this early stage of education, where the use of networked information sources and computer-mediated communication are of benefit to learners, but there needs to be a thorough appraisal of any implementation in terms of its impact before it is used. At later stages of education these problems do not have such an impact but there still needs to be a recognition of the merits of whatever approaches are possible and a choice based on user requirements rather than technological possibility.

The issues that arise from the use of multimedia communication in education are quite broad and the finer details can be debated for many hours. However, the user and providers must decide on the following, at least:

- Why is it necessary to use the technology?
- Is the cost justified?
- Do the users require this type of access?
- Are the users (learners and teachers) equipped to use the facilities provided?
- What effects will it have on the learners, teachers, support workers?
- What effect will it have on other providers?
- Does the technology allow access from all the target group of users?

Consideration of these points should take place in the design stage of any technological approach to education, especially those based on multimedia communication networks. Resolution of these issues will only enhance the eventual process, even though it may be difficult at the time!

FURTHER READING

As the application of multimedia communication technology to education is a relatively new area most of the readings can be found in conference papers and proceedings. The IFIP TC3 series on teleteaching covers this area in depth and many of the studies referenced are taken from these conferences. The following are the most recent.

G. Davies and B. Samways (Eds.), "Proceedings of the Teleteaching '93 conference", Trondheim, Norway, 20-25 August 1993, *IFIP Transactions A-29*, North-Holland, Amsterdam, ISBN 0 444 81585 6.

J. D. Tinsley and T. J. van Weert (Eds.), "Proceedings of the sixth IFIP World Conference on Computers in Education", 1995 (WCCE '95), Birmingham, England, July 1995, Chapman and Hall, London, ISBN 0 412 62670 5.

W. Veen, B. Collis, P. de Vries and F. Vogelzang (Eds.), *Telematics in Education: The European Case*, Academic Book Centre, De Lier, Netherlands, 1994, ISBN 905478028 2.

Specific references

A. Sloane, "Multimedia and communications: A quart into a pint pot?", NDISD, March 1994, Wolverhampton. Paper available on the Web site associated with this book.

S. Coiro, F. Davoli, P. Maryni, P.P. Puliafito, M. A. Pavan, P. Talone, "Design and experimental monitoring of an ISDN-based multimedia distance learning service", *Proceedings of the IFIP TC3 Third Teleteaching conference*, Teleteaching '93, Trondheim , Norway, pp157-166.

L. Stefánsdóttir, "KIDLINK: Creating the Global Village", *Proceedings of the IFIP TC3 Third Teleteaching conference*, Teleteaching '93, Trondheim, Norway, pp837-844.

Other material

The Web site associated with this book also holds more information and further references and pointers to more information associated to the subject of this chapter.

EXERCISES

1. Consider any course that you are currently studying or have studied in the past. List the necessary information and systems that would be required to implement that course as a teleteaching course to a group of learners spread around the country.

2.   For the following list of subjects, at secondary school level, discuss the suitability and the needs of a networked approach to information and communication for a group of learners in a conventional school environment.

a) Geography
b) Mathematics
c) Information technology
d) Craft, design and technology

3.   List the advantages and disadvantages of learning a new language from either

a) a CD-ROM or
b) using a network with links to another country.

Decide how you would like to see a hybrid approach, using both networks and CD-ROM, implemented. List the features and media used on each type of approach and indicate their use.

4.   If you have access to a WWW browser try the following. Choose an educational subject of personal interest. Find some WWW sites related to the subject. Follow the links to more pages related to the subject and compile a table of the sources of the information, listing the source by country (us, uk, fr etc.), domain (edu or ac, com or co, org etc.) and number of pages on interest at the site.

THE DIGITAL LIBRARY

Summary: The library in the digital age. Digital information and conventional media. Access to information, "reference" and "lending". Information storage and retrieval systems for libraries. The distributed library. Effects of multimedia communication and networks. The consequences and possibilities for the digitisation of knowledge. Copyright and future possibilities.

7.1 THE LIBRARY AND DIGITAL COMMUNICATION

In general, most people would define a library as a collection of books. This is still a dictionary definition despite the steps libraries have taken to introduce information stored in forms other than on paper! The accepted view is that what libraries do is store books which contain information. The normal method of use of any library is either to borrow books from it or to study them within it. This allows anyone with access to the library to have access to the information contained within it. This model of access to information has been adequate while most information has been stored on physical media that are accessible by simple means such as books, tapes and records. The move from these traditional forms of information storage to an all-digital form while allowing easier access for computer users, denies access to those that cannot use them. This chapter looks at this possibility in depth and discusses some of the assumptions, problems and driving forces behind this digitisation of knowledge that is being predicted.

This book is a typical example of a traditional approach to information storage. The information stored on the pages includes both text, images, drawings, and some tables or structured information. There are many books similar to this in libraries around the world. There are also many instances of audio and video information stored in simple cassette forms which allow easy use by almost anyone with the right machine. So all media forms of information are currently accessible by many users using a variety of machines for access. The links between items are relatively difficult to control. For example, a book could have an audio cassette of information attached to it, but the points at which the reader needs to play the cassette need to be marked in the book and the book should be able to be understood without reference to the cassette.

Although a number of publications have tried to combine audio and video with text and pictures on different media they have not achieved a great degree of success. This is mainly due to the problems and difficulties in using the information sources together. A magazine may publish a cassette of music and sell the two items together but they do not generally have strong linkage and the two separate items can be successfully used alone. Digital forms of multimedia information are integrated and linkage between items can be stored in or with the media making them a more unified approach to information storage and representation.

The use of communication networks alters the picture further. Multimedia information sources (or separate media information sources linked together) can be accessed via communication links and the information stored in a library can be made available to many more users, than has been possible in the past. The possibility that access to the information can be simultaneous also arises in a digital scenario. The problem has changed from being one of storage of, and access to, books and other articles to one of storage of digital information and access to computers. There are, however, many other problems, that are common to both approaches, not least of which is the area of copyright and payment for authors (a subject not too far from the thoughts of any author!).

7.2 WHAT IS A LIBRARY?

The previous section gave a traditional definition of a library. As was subsequently seen, there is much more to a well-organised library than just books. The information can be stored in many forms including audio, video and some mixed media formats such as audio-visual combinations. There are also many different types of library serving different communities and functions, some provided via public taxes, some via private or commercial funds and some being provided for specific groups. Many of these have different functions and some are more amenable to a digital approach than others. The main types of library are all described in the following paragraphs.

Lending library. A lending library is a common occurrence in many cities and towns around the world. Originally designed to be an adjunct to education for the masses they became common in the 19th century with mass migration into towns in western Europe and elsewhere. The original purpose has largely been replaced with a more leisure-oriented role in the late 20th century. The lending library is now more likely to be the source for the latest blockbuster fiction or spy novel than of a serious academic work. The size and scope of the information stored in local lending libraries varies widely in response to the size and complexion of the community that is served. Some town or city libraries contain many thousands of volumes some of which may rarely be seen elsewhere, whereas local community libraries often have little more than the most popular works available for lending. Mostly the stock of books in a local library is tailored by the needs of the lenders with more obscure tastes being satisfied by inter-branch lending. Some lending libraries keep other more specialised collections, often of local significance, which are regarded as national centres of their subject. There may also be an element of non-book information stored at any of these lending libraries. Some larger ones have complete audio and video collections and lend these to borrowers in the same way as books. The main difference between a lending library and a reference library is the borrowing of material for use elsewhere.

Reference library. There is, however, still a role for local libraries in providing information, especially concerning local events and issues. This leads to the concept of a reference library. The use of specific material for reference purposes is allied to the educational role of libraries, where material has been stored for scholars and other interested persons for many years. There are now many purposes for a local reference library that cover areas such as: general advice information, national and local information campaigns, information about local organisations and meetings, technical and safety information etc., along with the more traditional information provided by reference works and more specialised publications. The reference library is a common sight in large towns and cities, often combined with a lending library it may serve a much larger area than the lending library as it provides more specialised information. The reference library usually has a policy of access that allows greater use of information than is possible in a lending library since the information is retained in the library rather than being borrowed. This has the effect of forcing its use to be confined to the library and access to the information is restricted to those who can access the library itself. Generally a local reference library performs a similar function to the next category, the academic library, but to a different audience and for a different purpose.

Academic library. This is a more specialised form of both reference and lending library which is usually attached to an academic institution such as a university. The role of the library is to provide information and serve the needs of the user community. In the university case the library has a range of users stretching from undergraduates to academic staff where the information needs are, for undergraduates, largely limited to course-related texts and a limited number of academic journals and, for the staff, mainly related to research interests and highly specialised. So, although the topics of interest may be limited in any one library, the range of material can encompass the whole spread of knowledge from the general introduction to the highly specialised academic journals that are now published. There are some advantages for library management in having such a well-defined community. It allows much wider consultation between book and journal purchasers and the user community than is possible in a more public environment. There is, however, inevitably a financial constraint imposed upon most libraries in academic institutions that do not allow a full range of provision to be made for all the users' requirements. The twin problems of ever-greater specialisation and increasing financial constraints have made academic libraries a major target of research in digital techniques. This is already apparent in some areas where journal article abstracts are searchable on CD-ROM, journal contents are sent via email to interested users and some publications are now only available in an electronic form. In time a full digital library should be available along the lines suggested later in this chapter.

Company library. Mostly, any library within a company or other organisation may be considered to be a specialised reference library for the use of a particular group of people, i.e. the employees of the company. This type of library will inevitably be specialised and relatively small, although on its own subject it may have a large collection compared with a similar academic library with material on the same subject. The specialisation of such a library will depend on the nature of the company's business, but in a highly technically-oriented company the information used may encompass nearly everything that is known about the area associated with the company's products. Information may be stored in a number of formats and in a non-conventional manner with no physical library being in existence. This will not, necessarily cause any problems unless information is not available at the required time and place. The range of formats and media for information of this type may present some problems especially if the company does not have a strict policy to deal with information. An ad hoc library and information policy may result in loss of information and wasted resources while staff hunt information in unordered documents and books etc. The company library may be the most likely to gain from a digitisation of information as access will be easier, assuming the company is already computerised.

Children's library. A specialist version of the lending library for children is often made available separately. This allows specialist staff to concentrate on the requirements of the younger borrower who is most likely to require text geared to their reading ability. There will also be a place in a children's library for the other media such as audio and video especially for the non-reader or children at the pre-reading stage. Often the combination of media used in children's books varies more than in general books with a higher proportion of images used (picture books). This has always been used as an aid to reading and any digital version would need to deal with the extra data required to store the information in the picture elements of any book or document. The main problems are similar to any other library with the added dimension of the need to store and retrieve information appropriate to the reader's ability to read. In a multimedia environment this could be a tailorable variable which is determined at the time of reading.

Audio library. Some local lending libraries offer audio as a separate entity in a specialised audio section or separate library. There also some more specialised audio libraries that service industries that rely on sound, such as radio broadcasting, advertising and the recording industry. All of these face some similar and some particular problems when dealing with audio. In the general lending library the cataloguing of information has long been done for books and so on. Audio material is more difficult to classify at times and reference to it is not always easy. This problem is even more acute for the specialist audio library where individual sound snippets of audio information need to be accessed separately. A database linked to the sound material is one possible solution. Digital information is more amenable to linking into a database. There are also problems with the variety of different audio formats that have been used over the years with records of at least four different speeds, reel-to-reel tapes at a similar number of different speeds, cassettes, cartridges, CD and now a number of different digital formats such DAT (Digital Audio Tape).

Video library. Video libraries suffer from similar problems to audio libraries. There are also as many variations of library as in the audio situation, from local lending libraries (including commercial outlets) to specialised commercial video libraries serving the needs of the television industry, or some other video-related company. Much of the information stored in video form is stored on a variety of cassette formats. In the consumer market this has stabilised to the VHS format, which is now nearly universal in its application. The formats for professional and semi-professional video are not the same and require different playback machinery. The range of subject is as wide as in the audio case and special video collection databases can help in the cataloguing of information stored in video form.

Picture library. Finally, a more specialised form of library than most of the others mentioned so far. The picture library has been associated with the publishing industry for a long time. Many individual libraries exist to serve the needs of pictures users such as newspapers and magazines. The value of a picture library lies in the classification scheme used to access the required picture. The object is to find some image that relates to a particular subject, this may be directly related (the subject of a story) or it may be illustrative (an image of something related). So images need careful classification to allow retrieval in a number of different categories. Again, a database would solve the problem, but its design may be more difficult than is first imagined.

With all these different types of library there are a number of common features that allow users to obtain information. Three features that are fairly common are *access control, lending control and information classification.*

Access control. In the case of all the libraries there is some degree of access control. In a reference library this may be a reader's ticket that entitles the user to use the facilities. Some of the larger, more comprehensive libraries insist on a reader having a good reason for reference to the material contained in the library. This is mainly to deter the frivolous use of the facilities. Mostly the reference facilities in local libraries are provided by local taxation and are essentially free at the point of use, with no strict controls being imposed. Some more specialised libraries make stricter rules. Academic libraries are generally available to any student or researcher, but mostly their use is confined to those studying at the associated institution. Company libraries are necessarily more restrictive, in that they may contain company-sensitive information which is generally not available to the public. Some of the specialised libraries control access by charging for the use of their facilities.

Lending control. There will also be controls on the borrowing of material from a lending library. This may be by restricting the membership of the library, or charging for lending in the case of video libraries. Also a restriction on the amount of information borrowed at any one time is usually imposed, again by a ticket system. Any library which lends its stock to borrowers will need some system for retrieving the stock after use, and any subsequent charging that needs to be done for overdue items.

Information classification. The most useful of any reference tool in any library is the catalogue. This holds entries for all the books held in the stock and includes information to allow location of the material. The catalogue is closely related to the classification scheme used. A commonly used scheme is the Dewey decimal classification scheme which gives each book or item a subject-related class number. This is determined by the content of the book and ensures that similar books will be classified with a similar number. This number is then used when the items are stored on library shelves and is also stored in the catalogue entry. So any book or other item can be quickly found by using a catalogue which cross-references the title, author, subject etc. to the class number. For example, a well-known book on computer communications has a class number of 004.6 this would then be stored on a shelf of all the books with the same class number (usually in alphabetical order of author's name). The catalogue entry, either on a computer or on a traditional card system would have the rest of the book's information. In this case:

Author	Sloane, Andy
Title	Computer Communications: Principles and Business Applications
ISBN	0-07-707882-5

Publisher	McGraw-Hill, London
Year	1994
	278 pages
Class	004.6

This system for information storage and retrieval has been used for many years and has provided a good basis for many libraries. Any change in the system requires a lot of work in existing libraries, updating current stock and re-shelving material. It is, therefore, not a frequently performed activity. On the other hand, the computerisation of library catalogues has opened up a number of possibilities for users. It is now common to find academic libraries with on-line catalogues, and also many local lending and reference libraries also use computers to catalogue and track stock. Many of the systems in use are highly automated requiring only bar-codes to be scanned for any item to be issued to a borrower. In specialised libraries the database of information stored needs to have more structures to relate items to users' needs. For example, in a picture library the photograph in Figure 4.2 could be classified under any of the following headings.

Landscape
Norway
Cliffs
North Cape
Arctic region
Tourist attraction
18 July 1995
Photographer - A. Sloane

and probably some others. This type of information is essential for the effective running of this specialised type of information library but requires a lot of initial analysis of the information stored in the library.

7.3 MEDIA TYPES AND USE

In a conventional library there are a number of different formats used for storing information. The form of the information and the storage mechanism are all commonly used and well-known and will only be covered briefly in the following sections. The various media and the types of device (if any) used to retrieve the information are found in most of the types of library mentioned. Some, however, may be more likely to occur in specialist libraries than in more general reference or lending libraries. For example, databases of information are normally reserved for special purposes such as legal information which would be held in legal firms or professional legal societies libraries. The majority of information is stored in the conventional media of books, audio and video formats and picture images. These will form the main part of the discussion.

7.3.1 Traditional media types and use

In the context of a library, the media types covered in Chapters 2 and 3 occur in standard forms and in some common combinations. Mostly the combinations will be covered in section 7.3.2, but the use of image to illustrate will be considered with text in this section as it is so commonly used. The main types are text, image, audio, video and structured or database information. Although the different types of information and media are dealt with separately here a common approach is to use a unified approach to information classification. This allows a library to contain information of any of the different types listed here. In the following sections the items are considered separately to allow easier explanation. The probability is that most libraries will contain some or all of the types discussed.

Text. Text and the special case of text with illustrations is still the commonest form of information stored in libraries, whether reference, lending academic or some other library. This has long been the format used to store and communicate information and is still widely used. Books, pamphlets, newspapers, newsletters etc. are all mainly composed of text and this makes it the primary medium for communication between people when storage is considered. It has a long history of use and the technology of books and printed material dates back many centuries and pre-dates the other technologies (except for primitive pictures). Books and other textual information is still widely used and their competent use is seen as a prerequisite for education and learning in most countries of the world. The use of image to illustrate text has also been widely used for many centuries. Some of the earliest texts included illustrations and it is still a common tool to aid understanding of the written word. The mainstream systems in libraries currently allow text to be accessed easily and are mainly geared to the storage and retrieval of knowledge in the predominant form of the book. Access to individual parts of books is more difficult as the level of information does not primarily extend to this level. To access a book on a particular subject is relatively straightforward, the catalogue assists to this level, but to know whether a particular book contains a section on a particular topic requires access to the book itself. As an example consider the information about the book in section 7.2. This shows the book contains information about computer communications, but to see if ISDN is covered requires the contents page, and to see if it covers GSM telephone systems requires the index! The granularity of information contained in the catalogue is not fine enough to search for information at these levels.

Image. Images present different problems to books and other text documents. In section 7.2 the classification of the photograph in Figure 4.2 was discussed. This led to any number of subject categories that could be used to describe the contents of the image. This is common with most images and will be used by any effective picture library to cross-reference the contents of the library. This is a more difficult task than the book classification of the conventional library but is necessary for the system to work. Mostly, the systems used for access to pictures are computerised and use database techniques to access the image information. The categories used in the example are only indicative many more could be used for any one photograph. Again the granularity of information will affect the usefulness of the catalogue. For example, if the picture in Figure 4.2 only used landscape and photographer as categories then this would require users to do more physical searching for images than the fuller list given in the example.

Audio. Although audio has been the main means of human communication for millennia, it was not until relatively recent times that a convenient recording and storage mechanism became available. After the invention of the first phonograph by Edison many other similar recording devices have been invented using similar principles and all being used to record sound. Modern devices are now mainly digital in nature although the domestic consumer market only started using CD in 1983. Since then the analogue versions of sound recording have made a remarkable disappearance, although analogue tape is still common. Quality sound recording is now the preserve of digital media. However, many historical recordings are still largely stored in analogue form and this will continue until they are transferred to digital media which may take some time. Sound libraries suffer the same classification problems as picture libraries, although they are generally used differently in most everyday situations. Specialised sound and audio library will need the same type of catalogue information that allows cross-referencing between pictures.

Video. The use of video as a storage medium is even more recent than for audio information. The video recorder did not become generally available, or affordable, until the early 1980s although professional formats had been around for a number of years by then. From this time, however, an enormous quantity of video information has been recorded and stored on cassette. The late 1980s saw the introduction of the affordable video camera/recorder or camcorder and this led to more possibilities for personal information being recorded and stored as video information. Video libraries have similar problems to the audio and image libraries. Classification is difficult beyond a simple subject, or author approach and specialist video libraries need to cross-reference information with higher granularity for easier use.

Structured or database information. The addition of database or structured information into the library area has been a very recent matter and comparatively rare, except for the catalogue. Many databases exist for specific purposes and some libraries have access to these for users. An example is the abstracts of scientific journal articles held on a CD-ROM database which is common in academic libraries.

A database of information in this form may have many uses. The determining factors will be the features available in the search tool supplied with the CD-ROM and the amount of information stored for each item. The difference between structured information and other media is the size of the individual unit of information, which is generally much smaller in this case. The provision of a good search tool then allows a large quantity of information to be searched for specific results. For example, if the term information quality is used for a search in the abstract database the thousands of entries yield a result of a few articles. This is a considerable saving on time for the user. The paper-based version, although yielding the same result, would take about a hundred times longer to perform.

Many libraries do not solely keep one type of information and most compromise between the complexity of the systems used to access the information and the ease of use. For example, even if a library keeps a number of audio cassettes it is unlikely to use a different system of cataloguing for the audio items alone unless the collection is substantial. More likely is a combined catalogue of all the material held in the library, of any type. What is also common is the provision of a linked series of information on different media. For example, a book with additional information in audio format. This is common with musical information.

7.3.2 Mixed media

In many libraries there are a number of commonly occurring combinations of different media types containing related information. Some of these have already been briefly discussed in section 3.3.1. This section looks at those that occur most commonly in a library situation. The use of illustrations in books has already been mentioned as being one mixed media type, others discussed here include all the main media formats covered in section 7.3.1.

Illustrated book. Many books, except fiction, contain illustrations. These serve to add value to the text in the book. The addition of images is now commonplace in books and unremarkable. The information content is less easy to quantify than for a simple text-only book, although in a digital form images can require much greater storage requirement than the text. The proportion of text to image is relatively high (generally greater than 75%).

Picture book. The opposite end of the book spectrum from an illustrated book is the annotated picture book. Often used for small children or for books of art works where the image is more important. The proportion of text to image will be relatively small (generally less than 25%).

Tape-slide programmes. The spread of cheap, simple tape recorders and slide projectors led to some linked devices designed to replay a tape and show slides in a linked programme. These devices made the production of tape and slide programmes relatively straightforward. The slides and tape being synchronised with pulses written onto the tape to control a change of the slide. The producer could time change of slide for the best fit with the tape commentary. These have now been largely superseded by video, which although not ideal for this purpose can be used in other situations which are not possible with tape-slide programmes.

Language tapes and courses. Language learning occupies a special place in its use of media, having been a common first application of most of the new media inventions. It is now common to see language learning courses using text, audio, video, image and any combination of these media types. It is not surprising that it is also now available on multimedia formats. Language learning is assisted by the use of different media. It is far easier for most learners to learn language by associating sounds with images and most of the first language courses take advantage of this. Many lending libraries keep a number of mixed media language learning courses.

Audio-visual material. A more generic term used to describe mixed media use prior to the description multimedia is audio-visual material. This can encompass most of the media types discussed so far and was relatively common in academic institutions, although not universal. The term has been used to describe any combination of media in an analogue form which are used together to present a common set of information, such as a course of learning.

With the increased use of digital information the possible inclusion of all media types in one package became a reality. The CD-ROM allowed 650 Mb to be stored on a removable disc. It therefore became practical to incorporate text, image, audio and video in a digital form for playback using a computer. The initial use of multimedia was centred around devices such as these with different digital media types incorporated onto a single disc or source of information. As was seen in Chapter 6 this can be a little limiting. This is especially so in a reference or academic library context where information is required from numerous sources. The various digital forms are discussed in section 7.4.

7.4 DIGITAL INFORMATION

Digital versions of information have been covered in detail in Chapter 3. The treatment at that stage was necessarily generic with information types being considered rather than the digital forms of conventional information units, such as books. This section relates the earlier treatment to the commonly occurring information stored in libraries. All these types can be stored in digital form on a computer and this brings the benefits of digitisation to the information.

Books. A digital book would be a relatively small amount of digital information. An all-text book of average size, approximately 300 pages with 80 characters on 62 lines per page would require about 1.5 Mb of storage. The inclusion of each half page illustration would increase the total by 2 kb for a line drawing to 50 kb or more for an image. These limits on book size would largely depend on the number and quality of any images used, but should roughly fall between 1 Mb for a fairly small text book and 3 Mb for a well illustrated text (high quality art books are excluded).

Pictures. In a picture library the use of digital images would require a standard to be applied for the encoding. The size of the resulting digital versions of the images will then depend on the size and type (colour or monochrome) of the original image and the encoding method used to store it. Typical values could vary between a few kb for small monochrome images to many Mb for large colour photographs. Typically a small monochrome image of 15 × 10 cm scanned at 60 dots per cm using 256 point grey scale would require 540 kb of storage (uncompressed). A large 30 × 40 cm image scanned at 120 dots per cm using 24-bit colour would require 52 Mb (uncompressed).

Audio. The digitisation of audio has been widespread since the emergence of the CD audio standard. This format requires about 10 Mb per minute of high quality stereo audio. A lower quality sound standard could give acceptable results for as little as 0.5 Mb per minute. Most audio is packaged in CD-sized amounts of about 650 Mb. Any audio storage would require a huge amount of storage capacity to equal the amount stored on just a few CDs.

Video. Finally, video is an even more data-intensive medium than any of the others and requires many megabytes of storage. Typically, compressed video takes up as much room as uncompressed high quality audio. (A movie of about an hour can be compressed onto a CD.) Uncompressed, video could use from a few Mb per second to hundreds of Mb per second depending on the colours sampling, number of pixels and the frame rate.

So, the digital versions of the typical library objects can use many Gigabytes of storage for very little information. A typical example would be useful to illustrate the amount of information actually stored in mixed-media library.

Example. Consider a library in a small town or suburb that keeps stocks of books, audio CDs and video cassettes. The library is only a small branch of the local library service. The layout is as illustrated in Figure 7.1.

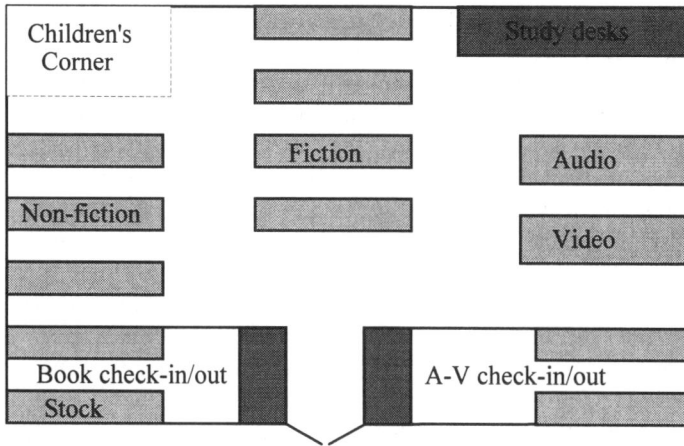

Figure 7.1 A branch library

In the library the total number of items held is as follows.

4000 non-fiction titles
5000 fiction titles
2000 children's titles
1000 audio CDs
800 VHS video cassettes (approximate length 2 hours each)
and about 200 different leaflets etc. containing local information, such as tourist attractions and concert series.

If this amount of information is to be held in digital form then the following rough estimate of the digital size of the information shows the kind of system that would be needed to store it.

Non-fiction	2 Mb per book	8 Gb
Fiction	1.5 Mb per book	7.5 Gb
Children's	1 Mb per book	2 Gb
Audio	650 Mb per CD	650 Gb
Video	1.3 Gb per cassette (compressed)	1040 Gb
Leaflets	50 Kb per leaflet	1 Mb

The figures show the relative scale of the information and large size of any storage that would be needed to hold mixed or multimedia information on a digital form. The main problem is the video information which is already compressed in the example. The other data could be reduced in size by suitable compression but the video information assumes an MPEG encoding. So a probable total for a typical small library would be in the order of 1500 Gb assuming compression was used throughout the different information types.

Using digital information in this way is unlikely. The wholesale conversion of libraries from a paper-based system to a digital format will involve both digitisation and re-organisation. For the main part there is unlikely to be any significant moves for this type of library as most users will not have the necessary access equipment, i.e. a computer. There will also be a long change-over period even if the systems do become available to access this amount of information efficiently and cost-effectively for unsophisticated computer users.

What is more likely is the partial provision of some specialised libraries, most probably academic libraries, and in some cases the pooling of library resources to access digital information held either centrally or in a distributed system between many libraries. The next questions to consider, therefore, are:

What are the benefits that digitisation of knowledge allows?
What are the problems with such a change to library systems?
How will it benefit the user and the library provider/employee?
What is the added value and are there any losses involved in this type of approach?

These four questions are discussed in the following sections.

What are the benefits that digitisation of knowledge allows? Information in a digital form is processable. In an analogue or paper-based form it is unlikely to be as useful. For example, the works of Shakespeare are much more use to the scholar in a digital form where words, phrases, characters and other pieces of information can be extracted and relationships established easily. In a book form the amount of work needed to achieve the same amount of cross-referencing and information extraction is quite large.

The ability to process information relies only on access to a suitable computer and the information itself. In analogue forms each medium needs special devices, equipment and knowledge to enable processing effectively. For example, audio requires tape recorders, filters, editing equipment etc. and video is even more complex. In the digital form editing can be done by software on a standard computer.

If a book is available in digital form the format of the index and contents pages can be different. The limited version of a book's contents (as included in this book) are a compromise between the effort involved in producing the list and the information required to be in it. It is possible to include more but the effort required to produce it is not worth expending. In a digital format the contents and index can be generated to suit the user's requirements.

A further possibility of using digital forms of books is the possibility of tailoring the book contents to the user's requirements. In the case of a children's book this could include tailoring the text to the age of reader. Having progressively more demanding text versions for different readers with a choice of different version according to reading ability.

Probably the most noticeable effect of digitisation of information is the ease of combination of different information sources into multimedia documents and related information. The whole process of multimedia has arisen because of the use of digital forms of information and this allows more types and more appropriate types of information to be used in any situation. These combinations are accessible and processable by computer. The whole information base available to any individual can be combined together to feed users' requirements. This inter-connectedness is a feature that is difficult to attain without digitisation of the information.

What are the problems with such a change to library systems? The use of digital forms of information will also cause some problems. There are a number of areas for concern. Firstly, the very nature of a library in a digital form is unclear. The purpose of digitisation is to render the different information types accessible by computer. They will need to be stored in a format that is usable by computer and this is often by storing them on a computer. The access to this information can be through communication channels rather than via a physical library building and the actual items of information need not be stored in as many locations as at present as communication access allows many fewer copies to be held. There is, however, a need for massive amounts of storage if all the data types are to be held in a digital form. This is especially true for video which is the most data-intensive of the media types.

The wholesale digitisation of all libraries is unlikely. Firstly, it would require too much financial investment for little return and secondly the users do not yet have the requirement for such a change. There may, however, be some small scale change in library provision in academic, specialist and company libraries where information is often required in different ways than in more public library situations. Academic requirements are often for individual pieces of information of a very specific nature. These can come from many sources and are currently available in books and academic journals. Digitisation will make their location easier. Using multimedia communication technology will enable them to have other media types incorporated and their use will require only access to computers and networks for the whole academic community to benefit from a single source of information.

This scenario does pose some problems for the current library organisations which are, generally, not geared to this type of information provision. In cases where information is available from single sources there will need to be a greater degree of co-operation between libraries in different institutions and countries than has been possible in the past. There will also be different skills required by the library staff who will need to use multimedia communications technology to interface with users, other libraries, and information sources. It is, however, unlikely that paper will totally disappear in the near future.

How will it benefit the user and the library provider/employee? The benefit to users will largely depend upon their needs. As was discussed in the last paragraph some users will find that information is more accessible via digital libraries but equally other users may find them a barrier to access. Non-computer users will still need to be able to access paper-based or other forms of the different media other than the computer-based digital format. So the main consideration in any change in the provision of information through a library will be the users' needs and their ability to access the forms of information provided. This will determine the point at which change may become desirable but not necessarily the point at which change takes place, which is more likely to be determined financially.

The library staff are also liable to find a large degree of change in their working environment. The use of computers to access information is already responsible for some small changes in library practices. If the total information stored is converted to a digital form then the current skills and practices of library workers will need to be used in a different way. There will be less contact with the physical resources of the library and more contact with the users and their information needs.

There will also be different opportunities for libraries to serve the users' needs. The current practice of on-line catalogues produces a record of the physical position or status of a book or other item. With digitisation the link between the catalogue and the information can be made complete and the software used for searching can include browsing the sources in addition to the display of titles as is most common at present.

What is the added value and are there any losses involved in this type of approach? Most of the advantages and disadvantages of the digital approach have been covered in the discussion of the first three questions. A summary is given in Table 7.1.

Table 7.1 Advantages and disadvantages of digitisation of libraries

Advantage	Disadvantage
Easy manipulation of information	Needs a computer
Better cross-referencing	Initially expensive
Access to more reference sources	Needs new user skills
Speed of access	Staff updating required
Multimedia approach	Old AV equipment obsolete
Can use a distributed approach	Books more difficult to read
Better access to obscure information	Libraries need to change

What is now clear is that there are a considerable number of issues that need to be considered before digitisation of information can be considered beneficial to all the interested parties, i.e. users and library staff. If a digital library is planned then the technical issues become important. These are the concern of section 7.4.1. The consideration of the user needs and social concerns is, however, not a peripheral concern but central to the direction of the technical research in this area.

7.4.1 Storage and retrieval requirements

There are many issues involved in the successful setting-up and operation of a digital library. Some of these are still the focus of extensive research in many countries and there are many possible solutions to the problems raised by the digitisation of massive amounts of information in a library setting. Some of these are of interest here and some are best dealt with in a more technical discussion of the relevant technological aspects of databases and communication technology. Since this book is concerned with multimedia communication the main area of interest will be in the area of access to the information stored in a digital library. To make any meaningful judgements both the storage and retrieval of information will also be discussed as will the various network tools and systems that are currently being used to perform some of the tasks associated with information retrieval and networked information provision, although a thorough treatment of HTML and the World Wide Web is left until Chapter 8.

In the calculations in section 7.4 the size of typical information objects and the different media types was outlined. If it is assumed that typical information objects will be related to those items in common usage at present this is a good estimate of the probable size of typical objects. However, having information available in a digital form allows more ways of using it than when it is presented in analogue or paper-based forms. This possibility can manifest itself differently for different media types, particularly in text, audio and video where the current paradigms can be altered for digital use.

Text. The common appearance of text is in the book. A book being a collection of chapters or topics which are related in some way to a central theme. In a digital environment the book could be replaced by a user-centred approach where different chapters of interest were each available separately via the network and if printed could be combined together. There is, however, no need to link information objects in such a system, except for the convenience of the user (who may want to take the paper version away and read it off-line).

Audio. A possible use of an audio library could be for compiling personal selections for recording onto CD. Recordable CDs are becoming much cheaper to buy and record and this could then be used to tailor the CD to the needs of the user. There may be artistic objections to such use of multimedia communication networks but the possibility is real. Instead of the user needing to buy someone else's selection or an album of tracks of little interest, the tailoring approach would allow personal CDs to be made in a relatively short time from networked sources. Charges for copying could be made via the downloading system in much the same way as information is sold on computer networks at present.

Video. The different types of video available are not amenable to different approaches. Films and fiction are necessarily stored in large blocks of information which require one to three hours to watch. There is, however, scope for more factual information to be combined into video sequences from different sources. The control over such a system would need to allow for different types of user since most users would not necessarily want to use video in this way. There are a number of different professions that might be tempted by such a possibility. Anyone needing to make presentations of material could use digital video in "snippets" to illustrate a lecture or talk. The video could be network-based and ease of use would depend on the ability to access it successfully. This would be largely dependent on the nature of the classification scheme used to access the material.

Initially the digitisation of information would depend on the form in which it exists at present. An evolution of the system will lead to new methods of working with digital information that may produce different approaches to information provision and packaging that will not be books, CDs or video cassettes.

The size of digital information objects will present some problems initially. In the example library in section 7.4 the total storage was much larger than would be affordable for such an application at present. Sharing of resources between branches and between groups of library providers would reduce the need to provide all the information types but there is still a need to store large amounts of information for users to benefit from a digital system. This then leads to questions about the access times and retrieval mechanisms used on such a massive system.

Access time. The time taken to access a particular item of information will depend on the storage technique used, the access mechanism or classification scheme, the database techniques used to locate the information, the size of the object and other factors. Some of these will be controllable by the user, some are decided at the design stage of the library system and some will depend on other users. Access time will be one of the crucial parameters determining acceptability of such a system.

Search and retrieval mechanisms. A further determinant in acceptability will be the ability to search the library, locate and retrieve useful information. This is a difficult area in which to satisfy the needs of users and the demands of library staff and management. The user wants to be able to find information quickly and easily using simple interfaces and commands they are familiar with. This requires a sophisticated search mechanism that needs each item to be meticulously cross-referenced and entered into the system. This, in turn, requires more work by the library staff to enter the information which increases the running costs of the library system.

Further complications arise if libraries are distributed between different systems. The access times will now depend on communication factors between systems and the search and retrieval mechanisms will need to have an awareness of the distributed nature of the information. There may also be the problem of cost tracking and accounting between systems which may belong to different companies or charging units. All of these factors increase the difficulty of the relatively simple idea of the digital library.

While the future shape of a digital library system may still be open to debate there are a number of subsystems available for use at present that allow network access to information resources. These have largely grown up around the Internet or have been fuelled by its growth. These different systems are essentially separate but could form part of future developments in digital libraries or may even disappear from the area of information provision. The use of email to distribute journal contents. the use of Internet browsing and fax delivery of journal articles, the use of CD-ROM and video on demand are all discussed in the following sections. The example of HTML documents and the World Wide Web is left for fuller discussion until Chapter 8.

Journal contents by email. One small initial step to the digital library is the increased use of email by academic journal publishers. The academic community is generally interested in the contents of journals and in particular articles, but to find out what the articles were has traditionally involved either subscribing to a contents journal which lists all the current contents of other journals or finding them in a library and scanning the contents pages. Both these options have flaws. Most academic libraries do not hold all journals and so important articles can be missed, and contents journals are an extra cost and often contain more journals' contents than are of interest to an individual researcher making the task more difficult. Use of email distribution lists can target contents lists more effectively. Each journal's contents page can be sent to interested readers immediately before publication and then articles can be noted for later reading. This makes the academic's task easier and is likely to lead to increased sales or use of the journals. It does not augur well for the producers of the contents journals since they could be redundant when all publications use email, but that is not imminent. The next step is to receive articles via email but it would require payment and there are a number of other issues that are relevant to such a system.

Journal articles by fax. One system that has been tried is the use of fax as a delivery mechanism for articles from academic journals. The user's access mechanism is via the Internet where they can browse abstracts and journal titles until finding something of interest. A credit card or account payment then allows the user to have a paper copy sent by fax. This provides the same facility as a traditional library journal facility that is commonly used by researchers and academic staff, either with locally held journals or via an inter-library loan facility with a central or national holding library. Such a system is, however, likely to remain only as an interim system until a more complete digital approach can be taken.

Electronic journals. It is, of course, possible to publish journals or books totally electronically, i.e. without an equivalent paper version being sold. This is now being done in a small way and can reduce the time taken between writing an article and it being published, which can be quite long in traditional academic journals. The review process, typesetting and production schedules can mean a delay of years between initial submission of an article and its eventual publication. In an electronic format the end product can be produced in a much shorter time without the need to wait for publication dates or by publishing more articles since the size of the journal will be less critical in an electronic format. Electronic journals are likely to form part of the future offering of digital libraries, especially in an academic setting where new information needs to be published as soon as is possible. For users later reference electronic journals could be distributed on CD-ROM.

CD-ROM. Since most multimedia applications and other multimedia information is now distributed on CD-ROM the digital library will need to allow for its inclusion in any networked approach. Reference to a CD-ROM is possible on a single computer or over a network. The problem is that a limited number of CD-ROM drives will be available and to service the needs of all the users all the CD-ROMs will need to be available all the time. To use a single computer requires the CD-ROM to be used exclusively in that computer and this does not allow sharing of the information. There are also licensing problems when software is used on a number of machines. Some use can be made of "jukebox" devices where a number of CD-ROMs can be held and retrieved as necessary and a number of these are now available for use on networks. These allow sharing of discs between users where there are a relatively small number of disc drives but the capability to use a larger number of discs. For example, a device could have four drives and space for 150 discs which can be inserted automatically when needed. This gives access to about 100 Gb of information. While this is useful in itself it also allows a number of different users access to different types of information at the same time. New standards for digital disc formats for larger scale multimedia information will only affect the amount of devices needed not the basic principle. In 1995 proposals for 4.5 Gb, 9 Gb and 18 Gb DVD (Digital Versatile Disc or Digital Video Disc) formats were published which should aid the introduction of digital libraries.

Video on demand. One of the most interesting developments in recent years has been the research based around the video on demand proposals. Essentially this is seen as a home entertainment service but the technological cross-over to the digital library is fairly straightforward. The main problems in delivering video on demand are the communication bandwidth requirements and the need for a server that can handle asynchronous requests to users. Some systems have already been tested and the idea seems to have a sound technological basis. The problems usually lie in the regulatory area. In a digital library the video needs are likely to be much less than in a video-on-demand system which is geared to the leisure market. It may, however, be seen as part of the portfolio of services offered by digital library service providers in the future.

7.5 NETWORK SUPPORT FOR DIGITAL INFORMATION

The main technological basis for digital library proposals is the ability to access information via multimedia communication networks. This has been discussed briefly in the preceding sections. Both the communication speed of a network and the ability of any servers to respond appropriately to requests for information are important considerations in the design of any network system that incorporates a digital library. In addition the relevant software to access the information in a manner appropriate to the users is also required.

Any digital library will need to hold a large amount of information of different media types. This information will need to be accessible in an appropriate time-scale. Most proposals imply that a document would be downloaded from a server almost instantaneously. This may not be the case. The communication speed of the network used to attach to the library system will determine the access speed. For immediate access and near real-time playback of video a high speed network connection with allotable bandwidth would be required. This may be possible in some academic institutions, research labs and companies but is unlikely to be the universal access mode for most users, at least initially. What is more likely is that access to information will be done in a similar way to the Internet, with no guaranteed data rates and mainly off-line use.

As an example, consider the use of a jukebox CD-ROM. The scheduling of requests needs to optimise the use of each disc access so that any queued requests for access to a loaded disc will be satisfied while the disc is loaded. However, to service all requests on a busy network will require disc changes to satisfy requests for other discs that may have been waiting. The algorithm should take into account both the waiting time and the currently loaded discs. This is a similar problem to many resource allocation problems that have been common in large computer systems for years. This scheduling algorithm will interact with the system to produce a delay dependent on the network traffic (in most current networks), the degree of use of the server and the overall design of the system e.g. the degree of dependence on the server as opposed to the client.

The raw data speed of the communication links used in such an application will have a large influence on the useability of the system. Based around current networks of shared 10 Mbps links text and image would be the workable limits with some low-quality audio. Video and high-quality sound would require too much capacity to be able to be carried reliably for more than a single user and that would interfere with the operation of the network if there were more than a small number of other users. Many libraries would, however, benefit from such a system to deliver information in these simple formats and many libraries would not require much more than this capability, at least initially.

A digital library upgrade path from current systems to high speed multimedia communication networks could start with text and image banks accessed locally, then increase the media types as the network is upgraded in speed. The upgrade path for wide area connections could use a similar argument with external sources being brought on-line as the local network is able to cope with the increased requirements of the information types. If the users are able to use non-urgent communication for information that is downloaded before use i.e. not real-time use then the problems are not so acute. However, the total data capacity required to support a large increase in the amount of video moved across a network is still beyond the current established LAN standards of 10-16 Mbps. The faster network standards offer an improvement, but for real-time access the use of dedicated channels or optimised video channels is preferable.

Any implementation of a digital library would need to be designed for optimum use of the network connections involved. Consider the academic campus in Figure 7.2. Here the various departments have their own library servers connected to a campus backbone network (which would be a high-speed fibre connection). Each individual server would specialise in the subject of its associated department and probably a particular information type.

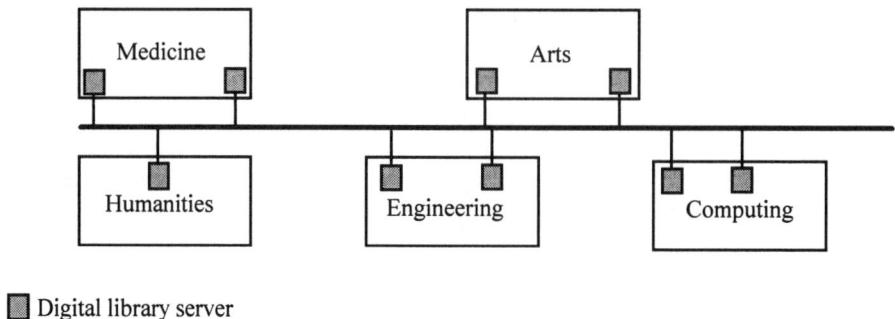

■ Digital library server

Figure 7.2 An academic campus digital library network

The user access is not shown but in the engineering department a user network would be able to access each server with access to the main library server from any station but some restrictions placed on access to the more data-intensive information held on the second server. In time, with suitable network upgrades this restriction could be eased so that all users could access any information of any type from any station on the network. Typical information types in each of the servers in Figure 7.2 could be:

Department	Server 1	Server 2
Medicine	Text and small image	Large image (X-ray, scan data etc.)
Arts	Text and small image	Large image, music, film
Engineering	Text and image	Design data, images, video
Computing	Text	Image, audio, video
Social science	Text and image	Audio, video

The exact division and proportion of each type in each server would need to be determined from prior usage and predicted access requirements of the users. Also, sufficient flexibility would need to be incorporated into such a system for change to be effected easily if requirements should change with use of the system.

This example is a simple demonstration of one possible distribution. Others are possible both on a single campus and where a single institution is located on a number of different campuses. In this latter case the scenario may be similar to that in the above example but with high speed LAN bridges between the different campuses.

On a wider scale a national library system would be likely to have a strategic plan for digital provision for an academic library service. One scenario could be as follows: a networked digital library based on a number of academic centres such as major universities each to cover a limited number of academic journals with local scope for others which are more popular. Books could be distributed in a similar way with reference text books used at any site stored locally and less used books stored at sites with major interests in the area of interest of the book. Other media types are more difficult to apportion. Network bandwidth would determine the number of copies required at various sites and the replication of data (of any type) would be determined by the ease of access and frequency of use. The initial allocation of the digital objects in the library would need to be done on previous use statistics of paper or analogue objects. Fine tuning could occur after the start-up phase of such a project. There are many questions and problems with any such system which can be solved by the technology. There are likely to be more questions from the human implementers of such a system that may not be concerned with the technical capability of the system.

One of the main questions to be answered in any implementation of a technological solution to any problem is that of access. In the case of the digital library this can be roughly translated into the following:

Who can access the system?
How is access controlled?
Who controls access?

In any particular implementation of digital library technology these questions will need different considerations. For example in the academic library system outlined in the previous paragraph the provision would need to consider the needs of the users at the various sites attached to the system so that the basic requirements of these are satisfied and that all allowed users can have the required access subject to any legitimate constraints (budgetary, security etc.). The control of access to such a system would need to be under the scrutiny of disinterested parties to ensure the fair distribution of resources both in the allocation of lucrative and important information and the setting up of connections to the network and the system. This latter consideration would require an assessment of the needs of those who are unable to connect initially for any reason (technical or financial) and those that may be disadvantaged by the effects such a system would have on the provision of information by other organisations. A reduction in paper-based journals would severely limit the effectiveness of organisations which are unable to connect to the digital library! Any access consideration would of necessity be a political question and any solution would eventually be based in the politics of the organisations in control of the implementation.

Finally, consideration needs to be made of the physical delivery systems for the different environments of the users in any digital library service. Most consideration is directed at the academic user in an office in a university as this is the likely first place of implementation of digital libraries. However, both the academic, commercial and home environments need to be considered for delivery of digital material. These different environments will have different characteristics in their information needs and the delivery systems will need to be designed to account for the different characteristics of the users and the groups of users expected of such systems. For example, the basic PC is envisaged as the main delivery vehicle in most academic and commercial environments, however, in the home different value systems apply and there may be more likelihood of TV set top boxes being used to access multimedia from a digital library in addition to video on demand. More details of the home environment are in Chapter 9. There are also special requirements in some industries that would not necessarily be satisfied with a simple PC delivery system. High-volume users of high-quality images and video such as the fashion industry or the digital film effects industry may need much higher specification equipment.

7.6 OTHER ISSUES

This final section looks at the broader issues associated with the digitisation of knowledge. Firstly a discussion on the important issue of copyright and the need for payment systems and then a look at some of the possibilities for publication in the near future.

7.6.1 Copyright

A definition of copyright states that it is the sole right to reproduce or record a work of literary, dramatic, musical or artistic nature. This now includes all forms of original information which have been produced by a writer or artist. The definition is necessarily broad to encompass works of an electronic nature, although not all countries have a copyright law that protects works solely in an electronic format. However, the intention of the paper-based copyright law which applies to books, music, images and films is clear. That is, that it is the right of the artist (unless they decide to forego that right) to exploit the copyright of a work. (Work is used in the broadest sense.) This means that it is generally illegal to copy whole works either by photocopying or other means.

The law is essential to protect the livelihood of professional writers and has been in place in various forms for many years. Similar laws relating to patent apply in other areas such as technology but for the manifestation of ideas and original writing the law of copyright applies to all works. The law is essential to protect copyright owners from unauthorised copying or other use of their works which would deprive them of income. The law gives to copyright owners the exclusive right to authorise exploitation of literary, dramatic, musical or artistic works as well as sound recordings, films, television and sound broadcasts, cable programmes and published editions.

If a work is copyright (some works are deliberately or implicitly put into the public domain and are therefore able to be freely copied) it should generate income every time it is used by a different reader or user. In the case of a book this is straightforward. A book is bought by a reader and a payment is made by the publisher to the author (eventually!). If the book is copied a similar payment should be made by the copier. If it is scanned and held on disc it should also generate a payment. If the book is stored on a server and a number of users can access the information the copyright holder has the right to payment from each user of the information. This is clearly a more difficult process to organise than the payment system for paper books or physical copies of pictures and films or videos.

In the UK, which has a modern copyright law, works attract copyright protection when they appear in material form — they do not have to be printed or published, an electronic form will suffice. No form of registration is required with a central authority. The owner of a copyright is able to profit individually from most primary rights that devolve from original copyright. However, secondary rights are more difficult to control, primarily as a result of modern technology or new rights, for example, photocopying, re-transmission by cable, off-air recording and private copying. These rights are difficult to administer individually and it is often more effective to administer them on a collective basis via an author's society or other collective body.

The protection of copyright is essential if the presumed growth of networks is to be anything more than an additive source of peripheral information. No professional author would allow a serious work to be sent around a network free of charge without some scheme of payment or acknowledgement for its origin. This problem has attracted a number of research projects in the 1990s as networks become more widespread and the digital library becomes more of a reality. The proposals are clearly aimed at providing a machine-based solution to the problem of copyright protection and payment for copying in a network-based system. There are, however, problems and requirements in any system of this type. These can be summarised as follows:

- Users must be aware they are accessing copyright material
- Users should be able to access material that is available
- Payments should reflect the use made of the information
- Payments should be related to the amount of information used
- Payments should be equitable

Obviously the first systems developed for copyright protection and administration may not satisfy all these criteria and many will not try but systems must be in place before the placement of income-generating works on networks can be seriously contemplated.

The systems in place for the payment of royalties to authors of all types of work are well established. When the works are translated into a digital form the system has no means (as yet) of generating the same information. Some of the proposals are based on audit-trails of information concerning access to original material and this can easily be linked to a personal identification number of a user to provide payment generation or invoicing. There are, however, still problems with secondary copying which has always been a problem for the publishing industry. New systems will need to be developed to incorporate copy protection in individual copies of copyright works.

Much of the work in this area is the result of current funding of research by government, particularly the European Union which has a large research programme concerned with information technology and telematics which has funded a number of projects in this area. Some up-to-date links to the results of these and other programmes can be found on the associated Web page for this book.

7.6.2 Publication for the future

Much of the discussion in this chapter has been based around the current paradigm of publication which is heavily influenced by the historical perspective of publication. The use of journals to publish research, books to provide a more thorough treatment and books to provide educational tools. Many of the first steps in electronic publishing have been directed at the previous methods of publication and the types of object familiar to users in typical libraries of any sort.

In previous sections of this chapter the publication of electronic journals has been covered and the type of multimedia publication available on CD-ROM. What is clear is that there are other opportunities when networks are fully integrated into the publication process. Without the need to put words onto paper there is a certain freedom of form that allows a "publication" to have more or less structure. For example, the use of hypertext is now seen as a way of providing information that can be used in a random access manner. This allows a number of "authors" to work together on a document or work, and link together parts of the work via hypertext links. (Chapter 8 covers hypertext in more depth.) This is a method of working that is difficult with the intervention of computers and computer-mediated communication which will allow on-line communication between the various authors as the work progresses. In such a system the amount of work submitted by each author may be relatively small. There is less constraint upon individual authors than under a paper-based system where, mostly, authors would work in chapters or sections. A hypertext document could be constructed from a thousand fragments of information each from a different author. The electronic storage of such a document would then be its publication. To a certain extent the World Wide Web is such a document although it is not all linked together as one document but as a series of separate, interlinked documents with a number of possible starting points.

Finally, the future of paper must be considered. The pressure on the environment to reduce the consumption of paper increases every year as does the publication of new works. There are, as has been seen, considerable difficulties in using networks for all publication purposes not least of which is the public resistance to the diminished access to information which would be caused. There are, however, more fundamental reasons for the continuing hybrid approach to information provision which are unlikely to change in the near future. The use of books and newspapers was widely predicted to be at an end when television was first used on a wide scale. The telephone was also predicted to have a deleterious effect on human communication. None of this has happened and the widespread use of networks is unlikely to change the human need for physical access mechanisms that are easy to use and widely understood. This is not to say that the use of networks will cease, far from it, they are set to grow and cover more of the world's population and allow access to more information than ever, but not as the exclusive source of information which would be to the detriment of much of the world's people.

FURTHER READING

Much of the subject matter in this chapter has been the subject of a number of research projects in recent years and there are many different sources of information available on-line. Mnay of these provide access to rapidly changing ideas and information. Reference to this book's associated Web page is seen as the appropriate way of providing references for this chapter.

There are a number of conferences dealing with some of the technical issues in digital libraries and some of the social issues are covered by the work of IFIP Technical Committee 9. Specific reference to these will also be found on the Web pages

EXERCISES

1.   Design a framework for the digitisation of a local video library. Use a local video hire shop as a suitable model for size, catchment area etc. Calculate the size of server requirements, type of access mechanism required and outline the scheme to be used for advertising titles, delivering requests and trailers and charging.

2.   For the example outlined in question 1, calculate what the current charges should be for a system based on the cost of equipment at the video library and the cost of films (assume the current average price of a cassette) to make it a profitable system. Assume that users are not required to buy any extra equipment.

3.   Using the current provision of books, journals and recordings in a public music library as a guide, design a multiserver solution to a digital library implementation. (Use a library with which you are familiar if possible.)

4.   Discuss possible hardware and software solutions to the provision of electronic novels. Compare your solution to the paperback book giving advantages and disadvantages of each approach.

HYPERTEXT COMMUNICATION

Summary: Hypertext as one of the main implementations of multimedia. Hypertext mark-up language. The World Wide Web. Information and distributed multimedia resources. Examples from business and education.

8.1 INTRODUCTION TO HYPERTEXT

The recent explosion in the use of the Internet and in particular the World Wide Web (WWW) has largely been fuelled by the use of hypertext, specifically the use of hypertext mark-up language or HTML. Before 1990 the major application of use on the Internet was electronic mail, a text-based service of use to the many users of the networks. When the hypertext transfer protocol was designed, it immediately allowed users to exchange other media types and to find information stored in different forms. This chapter aims to explain how this has become so useful and is the common face of multimedia communication in networks. Firstly, the difference between hypertext and HTML is discussed followed by a simple introduction to the use of HTML in WWW documents with a number of simple examples. Finally a number of example Web pages will be explored to discover how the WWW is being used for industry, education and personal use.

8.1.1 History

Plain text is, in essence, a sequential medium. A typical book is designed to be read from the front to the back. A particular case being a murder mystery novel where the plot is sequential and the outcome requires knowledge of the preceding points. Some text can be used differently. In some textbooks it is possible read in a random access fashion where individual chapters or sections can be read alone. A case of this is the encyclopaedia where each individual section is a stand-alone piece of information. It is possible to read an encyclopaedia from cover to cover, but not recommended! Many forms of using text have developed over the centuries since it was first possible to make books. What has changed recently is the use of new computer technology to store text and which has a processing capability which can be used to access text in a more random access method than was previously feasible.

Hypertext systems have been in existence since the late 1970s when computers first became popular for storing text. The ability to form links between parts of the document or between different documents allowed references to be included in text and the sequential nature to be removed. A document created on a hypertext system was able to reference and be referenced by any other document on the system. This allows new ways of constructing documents and text that were difficult to manage with a paper-based system. A typical sequential text could be denoted by the structure in Figure 8.1.

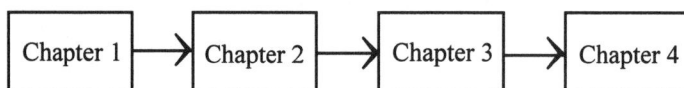

Figure 8.1 A sequential text

A document of this type does not require explicit rules for structure as each section follows logically from the previous one. However, in many books the implicit logical structure is more complex. Most non-fiction books have some amount of self-referencing (this book is a typical case) where parts of the text refer back and forward to other sections of the text and the reader can follow references as and when they wish. If a particular theme is of interest the reader can follow that instead of the strict sequence order of the text. This type of referencing is shown in Figure 8.2.

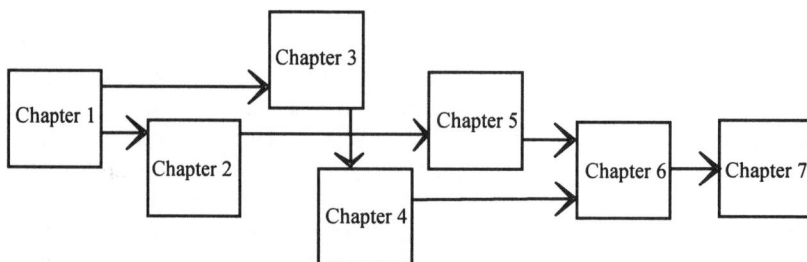

Figure 8.2 A linked self-referencing text

While many texts use this type of cross-referencing in the body of the text it is not explicitly part of the formatting of the document in most cases. Hypertext allows the inter-relationships between different parts of the text to be made clear and explicit. Each section can be stored as a separate object and the cross-referencing can be done via links between texts. There have been attempts to use such a linked system for novels where the outcome is determined by the path taken through the book, but these only had a mildly successful career. In non-fiction the use of hypertext is more obvious and in encyclopaedias it is almost necessary. It makes the use of reference material much easier.

It is only a short step between hypertext and hypermedia where the individual links refer to objects that are of different media types. This is, again, a consequence of the use of computers to access information. It has been tried in paper-based and analogue forms but requires a certain skill and tenacity to get the most out of it. The user needs to be able to access all the media replay devices, audio cassettes, books, and maybe pictures and video cassettes and then follow the sequence. The temptation is to finish the book and then see the video or vice versa! With a computer and the correct software all the different media types can be displayed on the same screen.

In all of these systems there is a need to be able to associate commands with the text or other media to allow the cross-referencing, i.e. to structure a hypertext document. This can take the form of embedded commands and its implementation will require some form of formatting language. The method used most widely now is HTML based on Standard Generalised Mark-up Language (SGML). There are other methods of achieving this objective but none have become as widely used. HTML is also relatively simple to use.

8.1.2 Relation between HTML and SGML

HTML is relatively simple in concept and has undergone a rapid development since first being used for formatting hypertext documents in the early 1990s. Whereas SGML is a complex general language that is able to be used to define and mark-up any document, HTML is an application of SGML and as such is very much simpler. The definition of HTML specifies the characters available for use and the way in which they can be used in an HTML document. This is contained in the Document Type Definition or DTD and the SGML declaration both of which can be found in the HTML definition document referenced at the end of this chapter. It is of little real interest to the user.

One of the problems with using HTML is the plethora of different versions that have been produced in a short time. The main variant, or common denominator is HTML version 2.0 which has achieved the status of an Internet standard. The majority of this chapter is concerned with HTML 2.0. There are also HTML 3.0 and HTML + which have arisen from extensions and upgrades to version 2.0. However, the simple applications used in this chapter can all be outlined and defined using mainly HTML version 2.0. Section 8.2.4 will give some of the details in these other versions.

To distinguish between versions users of HTML can include the document type identifier in an HTML document. A typical form of this would be as follows:

```
<!DOCTYPE HTML PUBLIC "-//IETF//DTD HTML 2.0//EN">
```

This would show that the document was prepared in accordance with HTML version 2.0 as defined by the IETF (Internet Engineering Task Force). All HTML documents should start with such a declaration although most of the browsers that read the documents can infer the contents by default if it is missing. The details of HTML are left until section 8.2 for clarification.

8.1.3 Use of hypertext

The World Wide Web did not come about by HTML alone, there were two other simultaneous developments that made it into the system it is today. These were the hypertext transfer protocol (http) and the use of the Uniform Resource Locator (URL). Http is the protocol used for the transfer of hypertext documents on the Internet and is not dealt with in this book. A technical description can be found in the relevant reference quoted at the end of this chapter. URLs are more interesting and are crucial to an understanding of how the WWW functions.

The terminology is a little confusing in that there are both Uniform Resource Identifiers (URIs) and URLs and sometimes they are referred to as *universal* rather than uniform. Either way it means a resource that is identifiable by a string such as

```
http://www.cs.anytown.ac.uk/home.html
```

This URL specifies a hypertext document named home.html on an Internet address known as www.cs.anytown.ac.uk. URLs allow specification of other documents in links between hypertext documents. So with the correct syntax a hypertext document can be linked to other resources on the Internet. These resources may be other hypertext documents or simple files available via ftp (file transfer protocol) and other resources available with other protocols.

So the three elements of http, HTML and URLs allow any resource on the Internet to be linked into a hypertext document. This is the WWW. These features, in turn, allow software to display the results of the transfer of information onto the screen. Because the protocol allows the specification of any resource a hypertext document can link to any media type making a hypermedia system. A hypertext document can be linked to the different media resources and played back on screen or through associated sound devices to enable a full multimedia communication system. This may sound like a solution to the problems of multimedia communication but as will be seen in the next section there are some problems with the approach as it currently works.

Regardless of these problems, there are many good features of the WWW that make it useful for both storing and retrieving information. In business, education and for home users the WWW has been the catalyst for the growth of use of the Internet. The ability to scan and browse hypertext documents allows users to find useful and related information about any subject that has been stored on computer. Best use can be made of the system when documents have been stored in HTML as this then allows hyperlinks between any related documents either on the same computer or elsewhere on the Internet. The use of HTML has made available many Gigabytes of information to Internet users that can be browsed fairly effortlessly with the various PC tools that are now available.

8.1.4 Problems in hyperspace

Before dealing with the specific details of HTML and its use on the WWW it is useful to consider some of the problems that can occur on a networked approach to multimedia information storage retrieval. This implementation of HTML on the WWW raises a number of issues when its use is considered. Amongst these are speed, navigation, appropriate links, media type usage, information availability and naming.

Speed. In general an Internet connection does not have a guaranteed speed. The use of the various links determines the capacity available to individual users. So even if a user has modem at 14 400 bps the actual data speed achieved will be determined by the various links between the user and the source of the information. When hyperlinks are followed around the WWW the source of information may be anywhere in the world and the speed of data transfer will be unpredictable and can often fall to unacceptable levels when accessing information from other countries. There is also a relationship between the media type used and the amount of data required to represent it as was outlined at the beginning of the book. The more data-intensive media types will, obviously, be slower to transfer. Any designer of hypertext documents needs to consider the user when including large data objects in any links. Fortunately, most of the advanced types can be disregarded if absolutely necessary.

Navigation. One problem with all hypertext and hypermedia systems is the way information is negotiated. The navigation through the links in hyperspace. This is particularly acute in the WWW since any link may be followed to a hypertext document that was designed by someone different from the designer of the originating document. The design rules may well be different and the use of hypertext links may also be different. So, a user branching out from a highly structured set of hypertext pages may be confused when confronted by less rigorously designed information. There can also be a problem when links are followed believing information to be available and finding it is not of interest. In this case the original document needs to be re-read and other links followed. While most browsing software helps with this it is not always easy to keep a mental picture of the information path as would be done with an encyclopaedia, for example. Some help to users from hypertext page designers is a useful aid to navigation.

Appropriate links. In addition to the problem of navigation through hyperspace the user also expects information to be appropriately linked together. Many users on the WWW have found that some pages have links to other pages of mainly unrelated information. This is partly caused by the race for publication that has taken place on the WWW where each user has produced information about all the subjects that interest them and linked them together, often with little regard for the design of the hypertext or the subjects linked together. Where this may be of interest to the original information provider it makes navigation and use of the information provided very tiresome.

Media type usage. The different media types outlined in the book so far are not all equally represented in the use of hypertext on the WWW. The fact that it is called hyper*text* indicates the bias towards text and its predominant position on the Web in HTML documents. This does not mean that other media are not used but that text is the link between objects of different types. There is some use of image as a link medium but the main agent is text. This bias is understandable given the capacity problems mentioned earlier. The use of hypervideo where on-line video clips can link to other media resources will require much greater guaranteed capacity before becoming usable on the Internet. These limitations on media type will need to be considered by any designer of hypertext documents for Internet use. The full use of all the different media in links together is a number of years off.

Information availability. One of the biggest problems faced by users is the lack of information about information. For the WWW to be of use in any situation the information required needs to be available i.e. it needs to be on a computer system somewhere on the Internet in a form usable by web browsers. This effectively means it must be in a hypertext format, which requires some work on existing information if it is to be made more user-friendly. The most significant problem is the lack of availability of information as providers (of all sorts from individuals to commercial service providers) do not always have the time or system capacity to make all information available even if that were desirable. More information does not always allow more to be used, it can confuse more than it helps. So information needs to be available and easily accessible.

Naming. Accessibility also depends on the naming of information. A filename is often used as an index to the information contained in the file. So for any information that is available it needs to be in a form that is recognised by the various search tools that are used on the Internet. Mostly this means using related file names that can indicate the content of the information. This has its limitations and does not always result in a search that gives adequate results. It may even give some information that is not relevant to the search but as the design of the systems stands at present there is little that can be done. A fully distributed database solution would require much more work than has so far been put into the WWW and that is considerable when the number of servers is taken into account.

8.2 HYPERTEXT MARK-UP LANGUAGE

Hypertext mark-up language is the foundation of the World Wide Web and any documents placed on the Web will need to be in HTML or linked to an HTML document for access via the hypertext transfer protocol. A hypertext document is essentially a text document with mark-up elements or tags which have a special format to distinguish them from ordinary text. The mark-up language is closely specified so that the various elements are defined for use by writers of HTML documents. What is now common is the use of an HTML editor which allows both text editing and the simple insertion of tags and other mark-up features. These editors are all helpful but not strictly necessary since all the mark-up features can be written in any simple text editor — it is just tedious to write out many of the features needed in a document.

The following sections are only a taste of HTML and are based on the most widely used standard HTML 2.0 and some on HTML 3.0. The level of HTML is intended to be basic and to allow further study from the relevant standards but the details should enable a number of hypertext documents to be written with links between them and allow the beginner to make a presence on the WWW. The various stages will be kept simple so that a number of the features of HTML can be explored and used in Web pages.

A fuller description of the elements of HTML 2.0 and 3.0 can be found in the book by Peter Flynn referenced at the end of this chapter.

8.2.1 Simple HTML

The structure of an HTML 2.0 document is relatively simple. The following HTML is typical.

```
<!DOCTYPE HTML PUBLIC "-//IETF//DTD HTML 2.0//EN">
<html>

<head>
<title>A hypertext page </title>
<head/>
```

```
<body>
<h1>Hypertext and the World Wide Web</h1>
<p> This document is a very simple HTML document to show the basic structure
</p>
</body>
</html>
```

This page, as it stands, doesn't do very much. It is simply a useful starting point for learning HTML. Firstly the various elements used will be discussed and then the likely output of a browser. The ability to accurately predict the screen display is not possible since the user may configure their browser to display HTML in a number of different ways.

The elements used. The document type declaration has been discussed earlier, but the structure is important. A document starts with a DOCTYPE declaration, then a head(er) and body contained within HTML tags. The tags used are in associated pairs so a start and end tag will always occur together in the same document around the features to which they apply. In the example the head is delimited by the <head>...</head> pair and the body by the <body>...</body> pair. The document is contained within a <html>...</html> pair.

Document headers. The header of an HTML document must contain a title, as shown between the <title>...</title> tag pair. It may also contain other information which is not part of the hypertext document as displayed when a user agent or browser displays the body text. The title may be displayed separately on screen and this is often the case.

Document body. The document body can take a number of different forms depending on the elements used in its description. In the simple example above the two elements used are a text section header as shown by the <h1>...</h1> tag pair and a marked paragraph as shown by the <p>...</p> pair. The text in the heading and the paragraph will be displayed in accordance with the settings of the user's browser. Section headings are normally used to display larger text as in the chapters of this book. There are six levels available h1 to h6. Paragraphs will be displayed in normal format as determined by the user's browser.

Display. The display is not fixed but one possible version could look like the display model in Figure 8.3. The text in the section heading format is shown in a larger typeface than the paragraph text which will be in the normal system font and size. The choice of sizes used in the display is largely up to the choice of the user and these can be easily changed in the common browsers available in all environments.

What this simple example lacks is a hyperlink to any other page so it cannot, strictly, be called a hypertext page as yet. The addition of links to other pages and

```
┌─────────────────────────────────────────┐
│               Browser                     │
├─────────────────────────────────────────┤
│                                           │
│   Hypertext and the World-Wide Web        │
│                                           │
│   This document is a very simple HTML document to show the basic structure │
│                                           │
│                                           │
│                                           │
│                                           │
└─────────────────────────────────────────┘
```

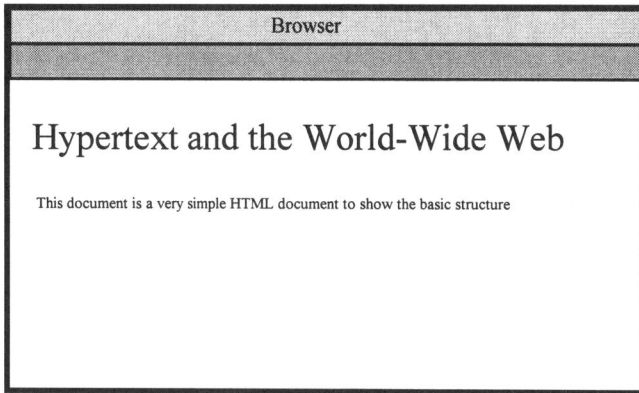

Figure 8.3 A possible display

media objects will make it into a simple hypertext page in a set of linked hypertext documents. This will require the use of media object identifiers which are known as URIs and discussed in section 8.2.2. There are, however, a number of other useful features of simple text mark-up that allow more structure to be used in documents. These will be briefly described here. The types of simple feature available are different types of list and other visual mark-up features.

Lists. There are a number different types of list that can be used. The main types are ordered lists where items are numbered and unordered lists where items can be in any order. An ordered list would be marked up with <ol>...</ol> tags and an unordered list with <ul>...</ul> tags. Each item in the list would have a list item tag pair around it i.e. <li>...</li>.

An example of an ordered list could be the different steps in a process such as dialling a telephone number. This could be marked up as follows.

```
<ol>
<li> Pick up handset </li>
<li> Wait for dial tone </li>
<li> If no dial tone - replace handset </li>
<li> If dial tone present dial number </li>
<li> If ringing wait for answer </li>
<li> If not ringing replace handset return to start </li>
<li> If not answered replace handset and try again later</li>
</ol>
```

The screen appearance would be as depicted in Figure 8.4 with the different list items numbered in sequence. This feature is also useful for any sequence of text that requires numbering such as agenda items, a list of lectures or a sequence of prioritised items.

```
┌──────────────────────────────────────────────────┐
│                      Browser                        │
├──────────────────────────────────────────────────┤
│                                                      │
├──────────────────────────────────────────────────┤
│  1. Pick up handset                                 │
│  2. Wait for dial tone                              │
│  3. If no dial tone - replace handset               │
│  4. If dialtone present dial number                 │
│  5. If ringing wait for answer                      │
│  6. If not ringing replace handset return to start  │
│  7. If not answered  replace handset and try again late │
│                                                      │
│                                                      │
│                                                      │
│                                                      │
└──────────────────────────────────────────────────┘
```

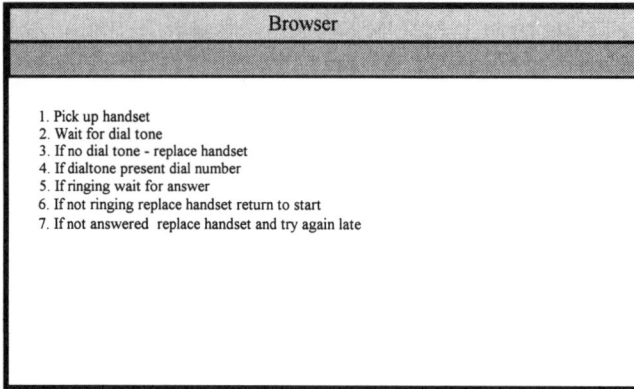

Figure 8.4 Display of an ordered list

An unordered list would be very similar except that the items would not be numbered but bulleted for display. So as an example of an unordered list the following mark-up would be typical for use in this situation. The various items are marked in the same way as those in the ordered list and the only difference is the <ul> tag.

```
<ul>
<li>Quad-speed CD-ROM drive</li>
<li>3.5 inch floppy disc drive </li>
<li>750 Mb hard disc </li>
<li>16 Mb RAM </li>
<li>17 inch colour monitor </li>
</ul>
```

The screen display would appear as that in Figure 8.5. There are other list types but these will be the most useful in preparing simple HTML pages and new users will probably not require any of the other features at this stage. One use of unordered lists is for the display of hyperlinks to other documents in a list format rather than in the normal flow of text. This will be demonstrated later.

Text mark-up. Other useful mark-up elements that are fairly simple are those that are used for the mark-up of text features such as emboldening, italics and underlining. There are also features for formatting text such as line breaks and horizontal rules. The following features are available in HTML 2.0.

...	Bold text
<i>...</i>	Italic text
<u>...</u>	Underlined text

```
                              Browser



        ●   Quad-speed CD-ROM drive
        ●   3.5 inch floppy disc drive
        ●   750 Mb hard disc
        ●   16 Mb RAM
        ●   17 inch colour monitor
```

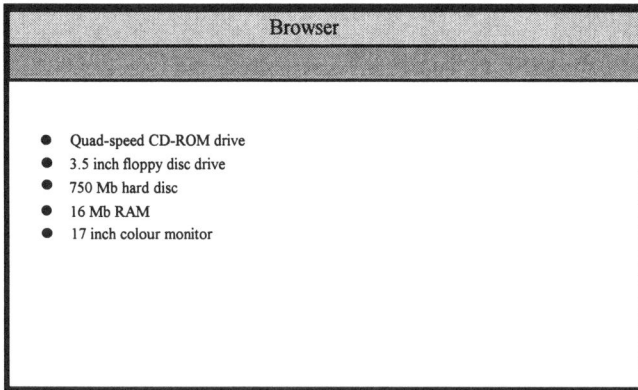

Figure 8.5 Display of an unordered list

` `	Forced line break (a new line is put into the text)
`<hr>`	Horizontal rule (a line across the screen)

The display of these features is reasonably obvious and will not be shown. There are also some other features which are unlikely to be as widely used as those already mentioned and these are left out to keep the discussion simple. The type of document that can be marked-up with the elements so far used is still limited but can include free text with lists and lines. The inclusion of extra features such as links to other documents is left to the next section. The last inline element that is commonly used is the inline image.

Images. The main distinguishing feature of a hypertext document apart from the links to other documents is the use of inline images to add graphical elements to the information displayed. This is relatively simple to achieve. The way HTML is designed uses images from separate sources (other files) in a document. Although links to other files are treated in the next section the use of images is slightly different in that they are displayed in the document as it is loaded into the browser. This is achieved by using a different mark-up to the one described later for external links.

Inline images use the `<img src=...>` tag. This can specify an image from any other accessible system and can use a full URL as described in the next section. If the image is stored in the same directory as the main HTML document it can use a shorthand version of the URL needed to access the image. There are also other attributes that can be specified. For example the simple mark-up line

```
<img src="picture.gif" alt="Text instead of picture">
```

places the image `"picture.gif"` into the document at the point in the text where it is specified and if the user cannot display graphics the words in the `alt` attribute are displayed instead. This allows all types of users to access the information in the document and does not disallow any users from accessing the HTML document, although non-graphic users will have a less interesting display even though the download will be quicker. In this case the image would need to be in the same directory as the main HTML document. There is also the possibility of using a graphic image as a map but this requires a knowledge of server script language and is too complex for treatment here. For information on this and other aspects of HTML the book by Peter Flynn referenced at the end of this chapter is a good source of help.

8.2.2 Linkage

As the main purpose of hypertext is to link documents together the foregoing mark-up only serves to format the document, the main ingredient is the hyperlink. This is the method used to reference documents to other documents within HTML and by using the Internet as a part of the World Wide Web. The WWW is a multitude of linked hypertext documents that refer to each other. Some are linked to many others some to only a few. The actual number of links is unimportant but the method of linking is crucial to the usefulness of the WWW as a tool for information retrieval and storage and as a business or educational resource.

The basis of linking any hypertext pages together is the use of URLs as described in section 8.1.3. This allows the specification of a resource for use by a particular protocol at a particular site, in other words it tells the browser where it is, what it is called and how to get it. URLs can be used in hypertext documents as links to any other accessible material on the WWW i.e. anywhere on the Internet where material is made available.

The basic use of a hyperlink is as an attribute of the anchor tag pair `<a>...</a>`. A simple example could be as follows

```
<a href= "http://www.cs.anytown.ac.uk/home.html">Return to Home Page</a>
```

This would link the document `home.html` on the server `www.cs.anytown.ac.uk` to the words `Return to Home Page` in the document. The actual style used for the display of the link is usually a change of colour, underlining or both, depending on the browser used. When the mouse pointer is moved over the link when using the browser the link name usually appears in a separate box on screen.

Hyperlinks to other documents are commonly used in HTML documents on the WWW. They can also be referenced to within the same document or to specific points in other documents. This is done by using labels within documents. A label is defined using another attribute of the anchor tag pair, the name attribute which defines the anchor as an endpoint for navigation rather than as a starting point. So the definition of the top of a document could be inserted by using

```
<a name="top_of_form">Some text</a>
```

and this would be referred to in a hyperlink by using the label. This can be done within the document by using the label alone as in

```
<a href="#top_of_form"> Go back to start of document</a>
```

or it can be used in a full hyperlink to another document by inserting the label after the resource name in the URL. For example,

```
<a href= "http://www.cs.anytown.ac.uk/home.html#top_of_form">Return to Top of
Home Page</a>
```

would link directly to the `top_of_form` label in the `home.html` document. (This is the implicit destination of the link without the label but the use of a label can be useful when different parts of document need to be referenced.)

There are a number of different hyperlinks that can be supported by most browsers. These use different protocols to transfer information back to the browser and are based on the various protocols in use on the Internet and WWW. Mostly the http protocol is used as this allows links to other hypertext documents and the WWW is mainly used for linking this type of document together but there are other protocols in use. These are telnet, ftp, gopher, news and mail. These can all be specified in the URL so that the browser uses the correct protocol to transfer information. For example, the URL

```
ftp://ftp.cs.anytown.ac.uk/results.dat
```

would point to the file `results.dat` and use the file transfer protocol to transfer it. The result of this transfer would depend on the set up of the browser for the handling of files. The most common browsers allow different file types, as identified by their extension, to be dealt with in different ways. The usual treatment being to store the file until it can be used later, although PostScript files may be sent to a viewer for on-screen viewing.

One other of these hyperlink types is the `mailto:` link which allows a mail item to be directed to the page author (or anyone else). If the link in the page is specified as

```
<a href="mailto:A.N.Other@mail.princetown.org.uk">Send mail to A.N.Other</a>
```

the browser allows the user to compose a mail item to the specified address. This is much more likely to illicit a response than simply putting the information in a text format in the document and eases the use of the mail protocol for an occasional user.

As mentioned above, the use of hyperlinks in documents allows links between any number of different resources. The simple hyperlinks demonstrated in this section are also mainly between separate documents but the inline images may also be linked from different sources, the simple example in the last section used the default of the same directory but it could have been any URL. For example an inline image could be linked with

```
<img  src="http://www.cs.anytown.ac.uk/picture.gif"  alt="Text  instead  of
picture">
```

linking explicitly to the source of the image. Users of this type of link would need to be careful when testing the page that the download to the browser was not unduly slowed down by the use of images from other sites and servers.

The use of most of the above features, both hyperlinks and the various mark-up elements, is demonstrated as far as is possible in the examples in section 8.2.3. Various examples will also be available at the associated on-line site which is linked to this book. The reference to the actual page is in the preface.

8.2.3 Sample HTML documents

As an example of a complete HTML document the following mark-up demonstrates most of the features so far described. The syntax of the various mark-up elements should be familiar so the following description will concentrate on the effect of the elements on the text of the document. The mark-up is as shown in Figure 8.6. The various elements are all as used in the previous section.

A possible appearance of the document is shown in Figure 8.7 but it must be remembered that browsers can be set up so that the various mark-up elements are displayed as the user wishes. The various elements are used to demonstrate the mark-up features and are not intended to be a finished document. However, the document as it stands would be able to be used and would provide both text and image inline and a number of links to other objects.

Text. The main text of the page is in a marked paragraph at the top of the page. This would be displayed as normal text except for those parts marked as being italic or bold. The various defined header styles are also displayed in sequence. These are also used throughout the document.

The other text mark-up features used are the line break and horizontal rule. If this is tried on a browser the text in the paragraph will not necessarily be the same length as that in Figure 8.7 since the window size of the browser and other factors will determine the actual size of the display.

Image. The image in the document needs to be present in the same file directory as the HTML document because there is no explicit reference to a URL so the default would be used. Another way to specify this is to use a head element called base. This allows the author of the HTML document to specify the default base URL to be used to resolve other partial URLs such as the one in this example. To use this a line such as

```
<base href="http://www.princetown.org.uk/a_n_other/">
```

could be included in the head part of the document. This makes sense when composing documents on one system for use on another.

```
<!DOCTYPE HTML PUBLIC "-//IETF//DTD HTML//EN">
<html>
<head><title> Multimedia Communications Test Page </title>
</head>
<body>
<h1>Multimedia Communication</h1>
<a name="top_of_doc"></a>
<hr>
<p>This page is intended to be used as a demonstration of the <i>various</i>
features used in <b>HTML</b>. The formatting of the text depends on the <i>User
agent</i> or <i>Browser</i> used to display it except that any explicit tags
used in the text will be used to add display character to the text. (e.g.
<b>bold</b>, <i>italic</i>, or <u>underline</u>)
</p>
<hr>
<h1>Using Headers</h1>
There are six levels of header style that can be used
<h1>Header Style 1 - Top level</h1>
<h2>Header Style 2</h2>
<h3>Header Style 3</h3>
<h4>Header Style 4</h4>
<h5>Header Style 5</h5>
<h6>Header Style 6</h6>
<br>
<h2>An image</h2>
<img src="pic.gif" alt ="A picture">
<br>
<h3>References to other Web pages</h3>
<br>
<a href="#top_of_doc">Go back to top of page</a>

<br>
<a href="http://www.princetown.org.uk/a_n_other/public.htm">
Author's Home Page</a>

<br>
<a href="mailto:A.N.Other@princetown.org.uk">
Communicate via email with the Author</a>

</body>

</html>
```

Figure 8.6 A sample HTML document

Multimedia Communication

This page is intended to be used as a demonstration of the *various* features used in **HTML**. The formatting of the text depends on the *User agent* or *Browser* used to display it except that any explicit tags used in the text will be used to add display character to the text. (e.g. **bold**, *italic*, or <u>underline</u>)

Using Headers
There are six levels of header style that can be used

Header Style 1 - Top level

Header Style 2

Header Style 3

Header Style 4

Header Style 5

Header Style 6

An image

References to other Web pages

<u>Go back to top of page</u>

<u>Author's Home Page</u>

<u>Communicate via email with the Author</u>

Figure 8.7 Possible display of mark-up

Links. The format of the links is again largely dependent on the browser and the display characteristics used. Most of the PC browsers use a colour change and/or underlining to emphasise a link to another document. Correct specification of the URL is up to the author as the document is not checked until a load is attempted. In the case of an inline this will be when the document is loaded into the browser but for an external link it will be when the link is used. If the user finds links that do not work a document soon loses its appeal!

All the features outlined so far are basic HTML mark-up elements and will be used in the majority of HTML documents. These are simply the links between text, graphics and other resources that users browse. To make HTML more interactively the Forms feature must be used. This is described in the next section.

8.2.4 Interacting with HTML - Forms

After using the WWW as an information resource it soon becomes clear that an interactive element is desirable. It enables users to send information to HTML document authors. The response data is composed of text and is often used for user questionnaires or other user surveys.

```
<!DOCTYPE HTML PUBLIC "-//IETF//DTD HTML //EN">
<html>
<head>
<title>A hypertext form</title>
<head/>

<body><h1>Fill in the following form</h1>
<form method="post" action="mailto:A.N.Other@princetown.org.uk">
<hr>
Enter your name<input type ="text" name="your name"
size=20 maxlength=40>
<hr>
Do you want to be sent a menu?
Yes<input name="Wants_menu" type="radio" value="yes">
No<input name="Wants_menu" type="radio" value="no">
<br>
<hr>
<p>
Choose your favourite food
<select name="Food">
<option>Fish & Chips
<option>Balti Vegetables
<option>Chicken Biryani
<option>Peperoni Pizza
</select>
</p>
<p>
<textarea name="Enter a brief comment"
rows="5" columns="20"></textarea><br>
<input type="submit" value="send form">
<input type="reset" value="Delete choices and start again">
</p>
</form>
</body>
</html>
```

Figure 8.8 A sample form

The basic use of form-based input to HTML documents is relatively simple to understand but the use of forms is more complex in that the information received from users will mostly need to be stored on the server and this will require a knowledge of server scripts. Again a reference to the book by Peter Flynn will be useful to any reader who requires this information. There is a simple option to enable HTML authors to test the output from forms. This is the use of the POST option with a `mailto:` URL which constructs a mail message to the author from the different input of the users. An example of this is given in Figure 8.8.

Constructing the HTML form is, however, as simple as constructing a normal HTML page. The various elements that can be used are mainly designed to make form filling easier. The main elements used are `<input>`, `<textarea>` and `<select>`. The **input** type is used to input single lines of text or to choose between different values of a set of options. The **textarea** type is used to allow input of a number of lines of text which can be specified in the form definition. The **select** type allows choice between different menu options.

The form marked-up in Figure 8.8 is a relatively trivial form which obtains information about food choices and the user's name but it illustrates most of the useful features of HTML forms. The three types of element are all represented in the example.

Firstly, the form requires user input in a text box that will be used for the user's name, then there is a radio button to press for a food menu to be sent, a choice from a menu of food options and finally a free text input area to make comments. The final two inputs allow submission of the form (construction of the mail message in this case) and a form reset to allow new values to be entered. A typical display is shown in Figure 8.9.

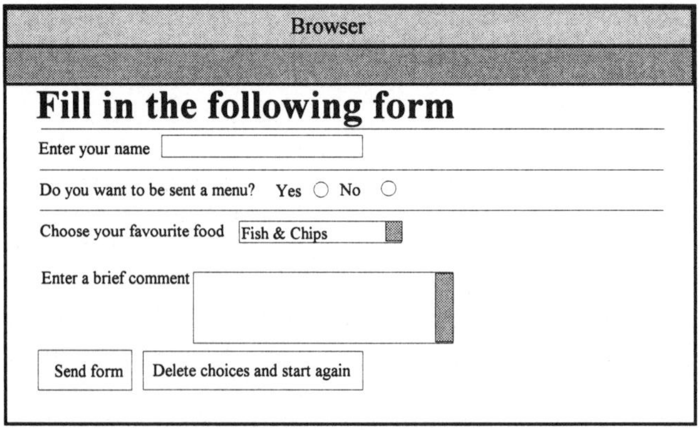

Figure 8.9 Typical display of a form

If the user enters the name A.N.Other and chooses Yes for the menu, Chicken Biryani for favourite food and please add more sauce as a comment the message that is sent would be as follows:

```
your+name=A.N.Other&Wants_menu=yes&Food=Chicken+Biryani&Enter+a+brief+comment=Please+add+more+sauce
```

the different fields are separated by the ampersand and spaces are replaced by + signs. This is not very useful as it stands, but its intended use is as an input to further processing such as storage on a database or as a trigger for more mail or a file transfer. Forms will continue to be a useful way of gathering information on the WWW as it involves less knowledge of different packages and procedures than using separate applications. Many companies and universities who use the WWW now use forms as a data gathering tool. Again for a more complete treatment the references should provide enough assistance.

8.2.5 Limitations and other HTML versions

Although this discussion has only covered basic elements of HTML there is little to be gained from repeating work available from elsewhere. What is clear is that the HTML standards will be developed as more users require more features for page mark-up and interaction. Any new developments will be publicised by the usual Internet method of "Requests for Comment" or RFCs or Internet drafts. These are published online and allow a discussion of issues surrounding new developments. They can be obtained from a number of sites around the world.

The process of updating HTML has been in operation for some time and there have been a number of versions available. The features outlined here have been mainly taken from the published version 2.0 standard but some of the newer version 3 features are interesting to note to see the direction of developments in HTML.

The most significant feature of text mark-up added to version 3.0 is the use of figures and the ability to define clickable regions inside them for the definition of hyperlinks. For example, an image can have a number of areas defined inside its boundary that a user can click on to access other resources. Although HTML 2.0 allows clickable maps it requires further server processing. The figures in HTML 3.0 allow the user to define links to areas of an image.

There are also additional mark-up features supporting mathematical symbols and the use of tables. The mathematical mark-up allows equations and formulae to be displayed in a realistic manner and the table mark-up groups items into rows and columns with headers and captions thus allowing much finer control over the display of information at the user's computer. For further reviews and more up-to-date information the reader is directed to the online information associated with this book. At present the limitations of HTML are mainly due to the limited number of features incorporated in its definition. These features will increase in time as new versions are defined and the limitations of HTML will become less.

8.3 THE WORLD WIDE WEB AND HTML

At the beginning of this chapter the use of HTML on the Internet was cited as being one of the main factors for the Internet's growth in the early 1990s. The use of HTML has taken the Internet from being primarily an educational and research network to being an all-purpose communication medium capable of being used for many purposes by all sorts of users. HTML itself has grown out of networks and networks have grown because of HTML in a cyclic dependency that is characteristic of useful ideas that co-exist. A similar phenomenon occurred between LAN technology and PCs each of which now depend largely on each other for continued growth and success.

As has been demonstrated in the earlier sections of this chapter the use of HTML allows users around a global network to communicate information to each other via linked documents and the whole becomes a source of information that is integrated and a world-wide database. This is, however, an ideal view; section 8.3.1 looks at how the Internet and HTML co-exist and section 8.3.2 will look at it more searchingly.

8.3.1 The Internet and HTML

The Internet was mainly used for a number of standard protocols such as SMTP email and FTP file transfer for many years. These were largely to serve the user community of the time i.e. research scientists and academics working on joint projects, sharing ideas and research results. This situation continued with relatively slow growth for about twenty years (1970-1990). Networks were interconnected and different countries were able to send information along inter-connected paths to exchange email and files and remotely log-in to other computers around the world. This collaborative approach to information also gave rise to HTML itself. It was originally designed to allow closer co-operation between scientists at the CERN research facility in Geneva, Switzerland and elsewhere. Soon after its initial release it became clear that the use of HTML would allow a world-wide information resource to be constructed by connecting information from different users together forming what is now known as the World Wide Web. To do this a new protocol, hypertext transfer protocol or http, was designed. This allowed hypertext documents to be linked together with the use of URLs as specified in HTML. The resulting links being resolved by simply clicking on an icon.

There are many reasons why HTML and the Web happened how and when it did. It is partially network related in that the Internet had the level of connectivity to support an application of this sort. It is partially related to developments in PC and other systems such as graphical user interfaces and higher speed processors. The combination of factors combined together to produce what is currently a usable system but one that will become even more useful when the networks are faster and information more easily accessible over larger distances. The need for faster networks being brought about by the requirement of users of this particular application to have better access to information! The situation is that the use of the Web has fuelled the growth of the Internet by adding more users; more users slow down the network demanding more capacity to give them better access, more capacity allows users to use more applications which need more capacity and so on....

Whatever the problems may be, it is clear that HTML and the Web will be a major use of the Internet for many years to come, especially as its full potential can only now be guessed at. With new features being added to HTML and improvements to the Internet being made, both the use and usefulness of HTML and the Web are liable to increase. The Web and HTML provides a simple interface for users to share information and to peruse the offerings of others. This application has brought the world of computer networks from the obscure areas of academic and research establishments into industry, commerce and the home. However, there are some drawbacks to this apparent universal accessibility.

8.3.2 Use and misuse

In the first instance it must be noted that the accessibility mentioned above only really applies in the developed western world and the Pacific rim countries where technology is used on an everyday basis. Vast portions of the world are under-connected and unlikely to be connected in the near future; they have more pressing problems to solve. However, even in those countries of the world which do have mass connectivity there are still portions of the population that are not able to take part in the WWW because they do not have access to the technology, either through lack of finance or education.

There are also significant sections of the WWW that are not used in a manner that would be deemed appropriate by all sections of the world community (it may be difficult to gain acceptance of anything on such a wide scale!). Many information providers do so because of interest or necessity, some do so for commercial considerations and some for political reasons. Whatever the initial reason information can have value and if the information is useful to a user, for whatever reason, it has served a purpose. However, there are times when information can have implications for the provider and the user that are not wholly beneficial. For example, there have been a number of stories in the press about the use of the Internet for both extreme political purposes and for pornographic material distribution. This is only to be expected. The Internet is a communication medium and will be used as other media have been used in the past. What it does provide is a much faster and potentially more widespread method of transferring such material between provider and users, often in a disguised or secret manner. What is difficult for governments to comprehend is the scale of the potential for this type of activity. A government is nation-based whereas the Internet is essentially global in outlook. The two do not correspond and any mechanism for regulation will not necessarily have the proposed effect.

Among WWW sites there are many varieties of outlook in design and information provided, some are good and usable some are less helpful. It is often also a subjective judgement. Many users like flashy graphics and links to other pages, some prefer a more ordered approach. Whatever a design achieves there is still a need to provide appropriate information for users. For a site based on interest this may be anything, but it is often helpful to users to limit the subject matter to a particular theme rather than construct a page that contains links to everything of interest to the provider. Commercial providers are mainly more focused in that they are often using the WWW service in order to promote their own products or services and this tends to increase the focus of the information in the pages. Other information providers are also likely to be focused on one particular area as the message tends to be blurred with too much extraneous detail.

For any WWW page design there are a number of factors that should be considered including: accuracy, appropriateness, layout of the information, copyright issues and technical considerations. Many pages, currently, do not take too much effort to ensure any of these matters. Information providers should take all steps to provide accurate information as there is no control over the subsequent use that may be made of any information provided via a WWW page. Appropriateness is more difficult to ascertain as the use of information, to a large extent, determines its appropriateness and this is, again, in the hands of the user. However, the information provider can ensure that all information is provided in the correct context and without too much extraneous detail to obscure real information. Layout of information on WWW pages can have a large effect on the use of information by users. Single screens are much easier to use and long documents do not help the browsing process. Too many hyperlinks are also difficult to use as information can become confused when repeatedly loading in new pages. Also, information providers should be aware of the copyright issue when using material from other people.

Finally, there are some technical considerations when compiling an HTML document. Including links in a page will inevitably mean that some users will follow them to the destination document. Page designers should be aware of the consequences of this action. If a page is provided on a slow speed link with a relatively low powered computer it may be counter-productive to direct a large number of users to that site as the capacity requirements will reduce the number of successful uses of the page.

8.4 APPLICATION TO PREVIOUS EXAMPLES

Most of the applications of the WWW are discussed in the various chapters that look at the individual application areas. The following examples are provided as WWW examples rather than as the more general examples of the use of networks and communications in the various areas covered in the other chapters.

8.4.1 Business on the WWW

The use of the WWW for business and commerce was soon initiated after the first WWW pages were sited in the early 1990s. The commercial potential of HTML was seen by a number of companies probably more so than earlier communication technologies such as electronic mail or file transfer. The reasons for this are linked to the external profile of WWW information rather than the inherently inward-looking approach of other information transfer tools. The WWW is a medium where users search for information rather than the earlier technologies where information was transferred between users. The WWW opened up the Internet to a new approach to information use and it is this that is the difference for business users. They can now provide information for browsing by users rather having to rely on users having an address to which information would be specifically sent. It is the equivalent to advertising as it has been practised for many years in various ways.

Internet and WWW business use is a growth area. The number of businesses now using the Internet and providing WWW pages is a significant proportion of all Internet sites. Some of these are little more than advertising material but some provide masses of information about products and other related areas. For example, there are now a number of computer-related companies that offer WWW sites to users where updated programs and information are stored for users to download to their own machines. Items such as new printer drivers are particularly helpful as these can be downloaded easily and quickly without the need to provide intermediate media, such as discs, for mailing. Many business sites also see the WWW as an opportunity to provide more information than is usually possible via traditional means. So it is common to find much more information on WWW pages than is often expected.

8.4.2 Hypertext as an educational tool

In an educational environment there has also been a considerable change in the use of networks and communications over recent years. While the WWW is not a significant teaching tool at present there is much potential for using HTML as a source for the information used in education. Many topics can benefit from a hypermedia approach where different media are used to best effect and where information can be stored in hypertext form for perusal by students. The use of browsers does not necessarily mean that information will be used in a particular way so there needs to be some control over the design of the information resources used for this type of information provision. However, with the increasing amount of information on the WWW there are a number of resources available to education that have not been available in the past. Items from business and other commercial sources can be used in addition to more traditional teaching material to give added immediacy to learning information and this can only benefit the students, as long as the principle aims are satisfied and the theme does not get obscured by particular examples.

8.4.3 Hypertext information resources

Finally, the use of hypertext as a general source of information is considered. In a library the information is sorted into categories where particular subjects are located together. This is designed to allow easy use and location of material and other items related to it. The hypertext approach in some ways mirrors this and in others is at odds with it. Hypertext allows links between items of a similar nature as in a library where similar books are shelved together, but it also allows links to be formed between items that are related but not similar. For example, a user may want to look into information about computer networks in a library. This may all be stored together in a computer networks area. If the research takes them into the area of electrical signals then this material is likely to be stored separately in an electrical engineering section. Hypertext documents and a mechanism for browsing such as the WWW will allow links to be formed between information of different subject areas as if it were all stored in the same area making its use much simpler for the user and allowing a more in-depth study from the material available.

8.5 FUTURE TRENDS AND POSSIBILITIES

The growth in the Internet has largely been as a result of the invention of the http protocol and the World Wide Web. This has enabled users to have access to information in a new way that is much more amenable to use than earlier protocols allowed. This has led to more information being made available than would have been thought possible until only recently. This in turn has led to the massive growth of both the Internet and the sites offering WWW pages for users to browse. This growth is both useful and problematic. The main benefit of the growth in WWW provision has been the amount of information from different sources that is now available to users of the WWW. The problems of the growth are caused by the fixed nature of the capacity of communication channels which, when used by more users, reduce the effective service given to all of them.

The problems may be solved by increasing the service capacity of the networks used to access the information held in WWW pages, creating a true information super-highway but this will take some time and any upgrade is not, at present, keeping pace with the growth in use of the systems and networks. When high-speed networks are available at a reasonable price there will be implications for the information provided via HTML. Currently, the amount of graphics and other high-data sources of information used on WWW pages is limited by users as the networks cannot transfer at sufficient speed to warrant their inclusion. This necessity to restrict media type use will disappear as communication links improve and users will be the first to benefit from the added information content available.

Also, further features in HTML are inevitable as it is still a relatively new technology. The simple mark-up will be supplemented by more advanced features and pages will become more complex and closer to the type of presentation that is currently the preserve of specific multimedia software, which is designed for individual platforms. This will tend to a blurring of the edges between local and remote resources making the tie-in to the WWW a more powerful tool than it is at present with individual presentations using both local and remote resources to provide educational, leisure and other material for users. The WWW is one of the tools that is likely to be around for a number of years but it is unlikely to be in its current form for long.

FURTHER READING

The first book includes a look at the background of hypertext and the origins of its use. As it was written before the explosion of the Internet and the WWW it lacks specific information on HTML.

R. Rada (1991), *HYPERTEXT: From text to expertext*, McGraw-Hill, ISBN 0-07-707401-7

A more modern text which discusses many of the same ideas is the following

J. Nielsen (1995), *Multimedia and hypertext: the Internet and beyond*, AP Professional, ISBN 0-12-518408-5

The third book is one of the best treatments of HTML for beginners and includes most of the features in HTML from simple mark-up to advanced use.

P. Flynn (1995), *The World Wide Web Handbook*, ITCP, ISBN 1-85032-205-8

Specific references

ISO 8879 (1986), Information Processing Text and Office Systems: Standard Generalised Mark-up Language, ISO, Geneva

The documentation for the specific standards of Hypertext Mark-up Language Specification - 2.0 and Hypertext Transfer Protocol (http) are the subject of further debate and can be found at the many sites that carry references to Internet standards. Further details are available on the Web pages associated with this book.

EXERCISES

1. Construct a personal WWW page using the simple HTML items outlined in the this chapter. Try to include at least one image and some references to interesting information from elsewhere on the WWW.

2. Conduct a survey of a particular subject. To do this choose a subject that is of interest, or possibly related to some work project or study topic, and use one of the available WWW search tools to locate information related to the subject. Make notes of the source of the information in terms of the type of site, where it was found (commercial, academic or other) and the country from which it originates.

3. Choose a subject from a local library and design a set of linked pages that specify the outline of the subject. Construct the outline pages and find some relevant links from local and other sources of information. (This question could be related to question 2 if a linked study is required.)

4. Design a set of WWW pages for your company or educational establishment to promote its activities.

9

THE HOME IN THE INFORMATION SOCIETY

Summary: The home as the next battleground for multimedia communication. When the "information superhighway" reaches the living-room what will be the consequences for the home-based user and what will be the benefits of having such capabilities. What systems are available for use and what are the systems being developed for the organisation of people and information.

9.1 INTRODUCTION

This book has taken a fairly traditional approach to the application areas of multimedia communication technologies in that their use is clearly a simple progression from current technology and practices. Business and education are often the first areas where new technology is applied and this has been the case with multimedia communication technology. In these earlier chapters the applications were mainly based around what is currently being proposed or achieved. In this chapter a slightly different approach will be taken. The home use of multimedia communication is still in its earliest stages and the coverage of current practice would be limited. This chapter, will, therefore, take a more forward looking approach to discuss more about the home in the future than about the current state of technology in the home. This is, however, not just idle speculation but a statement of what is possible, what is proposed and what may, or may not be wanted by the users. To enable any analysis of the likely impact of technology the current usage of information in the home needs to be analysed. Also the activities that take place in the home need to be considered to assess their interaction and use of information. First, though the home itself needs to be discussed since it, inevitably, means different things to different people.

9.2 WHAT IS THE HOME?

The home is a difficult concept to define accurately as people have different experiences on which to base their particular definitions. The next section will look at some of the common definitions in specific detail, but first the general concepts need to be discussed.

Historically, humans have always required a form of shelter in which to hide from the weather, wild animals or other humans. These spaces have been developed from caves and other structures into a range of different, and often elaborate, buildings. The building is only part of a home. The home consists of a living space and also other objects and people that are related to the particular person under consideration. By related it is not necessarily meant to imply that the people are related by blood or marriage as a home could consist of a group of individuals who share the same space. The concept of home is a strong one in western culture and it has been a developing definition as the centuries have passed. What would have satisfied a definition of a home in the 10th century would be far short in current definitions. A home is, therefore, a cultural object that is defined by what the indigenous culture considers to be a home and also by what the inhabitants of that specific home consider to be a home. A simple definition is not really possible except within the strict confines of a particular culture and a specific set of individuals that make up the home.

9.2.1 Specific definitions

To make some specific definitions of a home there are a number of assumptions that need to be made. Firstly, about the culture that is predominant in the area the home is situated and, secondly, about the individuals that inhabit the home. To analyse particular homes these parameters must be specified. If the home to be considered is one in the UK then the different cultural norms for different sub-groups of the UK population would need to be specified and any individual characteristics of the inhabitants would also be relevant to the analysis. There are, however, some common examples of homes within any particular culture and often the number of differences are less than the number of similarities between any two homes so some generalised analysis is possible. This may only be a general analysis but will help to elucidate on some common problems in the home.

As a working definition it is useful to assume that a home is a living space for one or more people where the inhabitants have a shared interest. To quote a more academic source a home can be defined as "the physical and symbolic frame and core of private life" (see the paper by Bjerg (1994) in Bjerg and Borreby, 1994, p. 329). Whatever formal definition is adopted the concept can only be defined within the cultural norms of the society of which the home is part.

In a western and/or northern European society, such as that in the UK, a home is generally accepted to be a house or flat (apartment) which houses a group of people. Mostly these people will be related i.e. they may be a family or they may choose to share the home and be unrelated. There are noted exceptions to this. For example, some people choose to live in hotels or in mobile structures such as caravans and they would define these structures to be home in the same way as a house-dweller. What is clear is that there is great diversity in the ways in which people relate to the culture in which they live and this can be manifested in their choice of a home structure for their own use. This is, however, fairly simplistic since not all people have a choice in what they can define as home and this can lead to many problems for them. To use these general definitions is relatively easy but it must be remembered, as the analysis proceeds, that the specific is not necessarily applicable to all possible home definitions and to make anything generally applicable would be nearly impossible. The simplifications made allow easier analysis and will be as general in outlook as possible. Firstly, the activities that take place in homes (within this restricted class of the definition) must be investigated.

9.2.2 Activities and interactions

Before the information used in the home and communicated to and from the home can be discussed the various activities and interactions that take place with and within a typical home will be described. These should be familiar to most readers although some may not have had much experience of all aspects of these interactions.

Societies have a number of different systems which interact to provide all the services and products used either in the home or by people associated with homes (e.g. in a leisure activity). Some of these are specifically related to the home and some are general systems that interact with the whole of society. A brief summary is given in Figure 9.1.

In Figure 9.1 the various systems that are shown each have a different effect on the home with different types and amounts of information flowing in and out of the home environment. Some systems may not interact with all homes, for example not everyone has an active participation in the education systems or uses the social services. However, this general view encompasses most of the sources of information that will directly affect what is done in the home.

There also considerable variation on the activities that individuals will carry out in a particular home. Even within a single home unit there is a wide variety of tasks and activities that are undertaken by the inhabitants. There is also a division of labour and tasks between the inhabitants that will determine who acts on certain information and, in some cases, what information is actually used in the home. The activities in Figure 9.1 may be split between all the inhabitants or there may be specific individuals who interact with one or more service and they are the sole point of contact. For example, a home may contain three persons who are unrelated and if one is a student

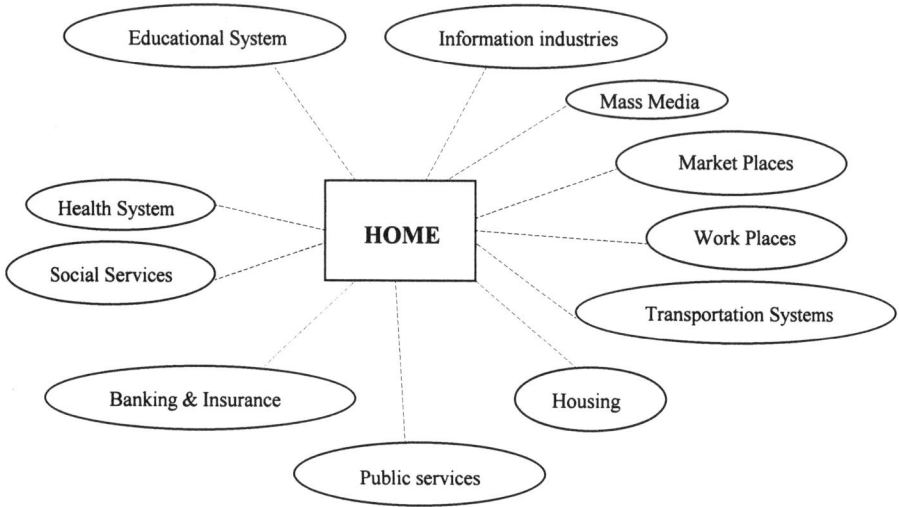

Figure 9.1 Interactions between the home and wider society

and the other two work for commercial organisations and do not currently use the education service then the sole interaction with education is the student, whereas the other two may be the only interaction with other work place organisations. They would all be likely to interact with most of the other services especially in areas such as transport or health.

What is also a factor in home organisation and information use is the activities that actually take place in the home. Mostly these activities are of a domestic nature and will be important information users, but there are individuals who use the home for other work-related activities that could better be described under a business heading. Many employees and a large proportion of self-employed personnel actually work in or from home making the home a dual-purpose centre. Systems for work and home have generally been different with home-oriented activities, devices and products being specifically targeted at the home user in all spheres of life, from cleaning fluid to information. When the two activities are co-existent the systems used have needed to be well-defined to enable the distinction to be drawn between home and work-related activity. There are instances, however, especially in the information technology area where the dividing line between work and home is very thin. The same space, the same computer and the same person combine to perform both work and leisure activity. This paradigm of working from home or in the home may have been used in various different jobs for many years but when the work is information-based and involves computer processing the distinction becomes blurred.

Of the other activities, most readers will be able to evaluate the effect they have on their own lives and the amount of information exchanged between the various systems and the home and its occupants. For example, the transportation system can interact in many ways, for car owners, various agencies or government departments are involved in driver and vehicle registration and licensing. The user will need to interact with insurance companies to exchange the necessary documents for vehicle and driver insurance and there may be other requirements depending on the country involved and the system used. For non-car users there may be interaction with train, bus and air passenger companies to exchange details of specific travel arrangements or of timetable information for regular journeys. There may also be user documentation to be exchanged such as season ticket purchase or similar travel schemes.

Most activities in the home have some reference to outside organisations and the flow of information is increasing all the time. The activities that are required for everyday life appear to be more complex and more demanding and the systems designed to interact with people in the home become more sophisticated. However, the basic activities remain as they have been for a number of years. i.e. mainly leisure-based with some domestic "house-keeping" included. The trend to increasing use of the home as a workplace will change this basic activity and produce different types of home environment. Some commentators see this as a major cultural change in that this may be the main paradigm in use in future years, as yet it is impossible to predict.

9.3 INFORMATION IN THE HOME

The activities mentioned in Section 9.2 all generate and consume information. Some are directly involved in information production and use such as the mass media and the education system, others do so as a peripheral activity to their basic function such as the example of the transportation system outlined in the previous section. All of this information needs to be processed by the home user or by a system for the home user so that they can gain the information that is required and provide the information required of them. To look at information use in the home the different types of activity need to be analysed and the different types of information determined along with the various media used to communicate them. This is a very complex task and each home will be different. The following sections will look at typical cases and uses of information in some homes and a general picture will emerge. At this stage the information considered will be of any type and based in any medium whether it uses electronic communication technology or not. So the range of information considered spans the whole sphere of home activity, from a note to the milkman (a common UK communication device) to satellite broadcasting.

After the discussion of what information is used the flow and use of the information is considered. As in a business environment, there is a degree of "supply chain" organisation required in homes with information being delivered through one medium and, perhaps retransmitted through another. For example, a person may receive information from various sources such as magazines, letters and television and produce a condensation of that information in the form of a newsletter that is sent to other interested individuals. This is often a paper-based exercise but there is scope for automation and computerisation of such activities to allow wider dissemination, better quality input and more timely production. This sort of activity is common in homes and is often disregarded by systems and communication providers as being of a low priority, but with increased availability and access to computers the home processing of information is a possible growth area in the future. This can also be said of the more financially necessary, work-related information activities that can be performed from home. To analyse the information that is needed by home-based users all the various sources of information need to be considered and the different formats that information is communicated into, and out of, the home.

9.3.1 Information and media types used in the home

The traditional sources of information used in the home are more diverse than in most other situations. This may not be apparent in any one particular home environment, but when all homes are considered together the mass of sources of information is staggering. These sources can be either formal or informal for any home. Some idea of the breadth of coverage is given in Table 9.1

Table 9.1 Traditional sources of information

Formal	Informal
Newspapers	Face-to-face conversations
Magazines	(with friends and neighbours)
Television and radio	
Letters (from companies or government)	Letters (from friends and family)
Newsletters (from societies or clubs)	
Telephone (from organisations)	Telephone conversations

The formal sources of information are both the mass media and simple one-to-one communication sources such as the telephone. The mass media, such as, newspapers provides one of the most obvious sources of information used by people in the home enabling the inhabitants to find out about various subjects of interest and providing other information that may or may not be required. This characteristic is typical of the mass media where information is chosen by an outside body or person, such as an editor or news reporter and then provided. Some information is not easily obtained and some that is not wanted is part of the total package. The traditional media do not have an easy way of making available small portions of information that are of particular interest to a single person. The scale of the mass media enterprise militates against such a possibility. Similarly the broadcast media are in a similar position although the user can choose whether to view or listen to particular offerings from each provider.

The one-to-one formal communication channels are often used for specific purposes. For example, companies may telephone clients to offer new products or check service levels. Organisations such as universities may use the telephone and letters to keep in touch with students on distance learning courses. Individuals may also receive communications from all sorts of national and local government departments which are specifically addressed to one person and so cannot be considered to be part of the "mass" media. These one-to-one communications are commonly the source of particularly useful information for individuals as they are directed at a single user and have direct relevance to that user.

All the different sources of information in Table 9.1 have common entry points into the home. The paper-based information sources such as newspapers and magazines are generally delivered or purchased; letters are delivered to the home (although delivery systems can vary). The broadcast media are delivered via specialised devices such as television sets and radio receivers, which enable users to receive particular types of signal. Newer signal sources such as satellite broadcasts require more equipment of a specialised nature especially if the subscription channels are required. The other main source of information is the telephone channel which permits voice communication channels into the home. These different sources and the entry points into the home are summarised in Figure 9.2.

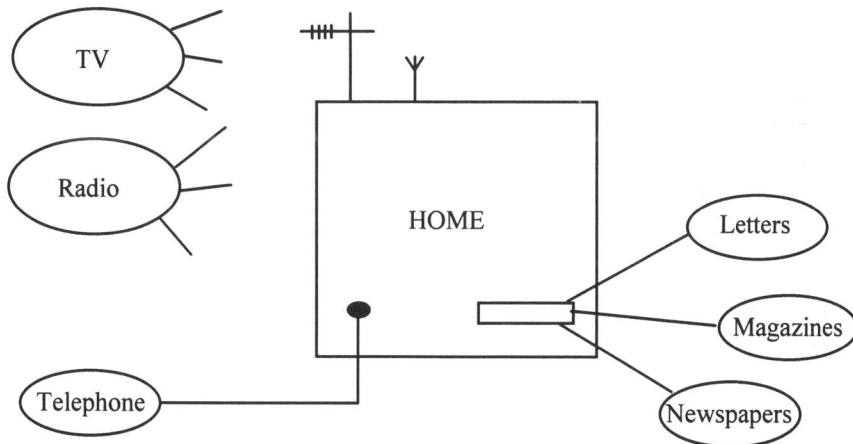

Figure 9.2 Information input to the home

All these different sources and entry points allow masses of information into the home in all the different media types that have become familiar through the rest of this book. These will be discussed in more detail later in this section. The entry points at present allow the different forms of information (different encodings) to be delivered to the home. These have been developed for the convenience of users over the many years since they were first used. People have also become much more sophisticated in their use of information. The variety of sources allows cross-referencing between different providers and the different entry points allow a redundancy of provision in case of difficulty with particular sources.

There are also many purposes for which information is used in the home. Some information is for leisure activities. Examples of this may be a list of concert dates mailed by a concert promoter. Some information is used for educational purposes, such as wildlife programmes on television (this is not necessarily formal education). Other information may be used for work-related activity, such as telephone calls from colleagues. Whatever the purpose the sources will remain the same and the entry points will be the initial point of access to the home. These points of access are changing with the new technology that is available with multimedia communication.

The types of information used in the home span the full range of information types and media used in other contexts. The difference is that often the information used in the home is less demanding, of a lower quality or less data-intensive. For example a video recording device used in the television industry is generally of much higher quality than the devices used in the home. This is generally so that the losses that occur with transmission can be reduced for users by using better source information. In other areas and with other media types there are corresponding levels of quality that are deemed appropriate to the home user. The types of data that are used can easily be related to the various sources of information that are used in the home. These are discussed in the following paragraphs.

Text. This has been the main source of information coming into the home until the 20th century when new technologies such as radio and television replaced text as a prime source of information input. Text is used in many of the sources of information, from newspapers and magazines to letters and notes, both formal and informal. There is still a degree of formality associated with receiving a letter that is not accorded the same information when using the telephone or a similar communication system.

Structured information. Although usually transferred to some other format such as text, the use of structured information is still common. Large database users often interface with the home user through the selection of information from a database. For example, a bank may want to target a particular campaign for savings accounts on customers who seem to have enough money to save. This can be done through the use of the structured information in the bank's computer. Similarly, home users can access the information held in the bank's database via a telephone banking service where a voice interface (a bank employee) relays the information to the customer.

Image. These are commonly used in a home environment and in many of the different sources of information that interact with the home. For example, images occur in newspapers and magazines to illustrate articles. There are also images that are used within the home in the form of photographs for use as records of activities and events and for an artistic display. Most homes will have some form of image even if it is simply the pattern on the wallpaper! Many homes will have far more than is realised and new images are fed into the home every day by daily newspapers and other paper-based information. The supply of images appears to be constantly increasing and is now used in a greater proportion to text in many publications than was previously possible using earlier technologies.

Audio. Since the invention of the phonograph and radio transmission the use of audio in the home has increased dramatically. Audio cassettes, CDs, and early technologies such as vinyl records all contain audio information. These sources supplement the daily input from radio transmissions, telephone conversations and less formal audio input, such as a conversation with a neighbour. All of these sources can contain audio information. The more formal formats containing information that is purchased and controlled by copyright. The less formal sources will depend on factors other than cost. Audio, being the primary medium for information transfer by humans

Video. Both television and recorded video cassettes now contain significant amounts of information for home use. Many homes rely on video format information for much of the information input to the home and there is no shortage of input. There are a number of terrestrial television stations in most countries which broadcast video information to homes within the range of the transmitters. There are also satellite television broadcasts which can be received by a number of countries within the "footprint" of the satellite transmitter. Together, terrestrial and satellite broadcasts provide many different sources of information into the home. This may be information for leisure activities, of an educational nature or for political or public service purposes.

All these types of media are common in the home and are commonly used to obtain information or to interface with information sources. They are also used to produce information which is transmitted from the home into other homes, into businesses and other agencies. Traditionally this type of home-based processing has been difficult and limited, but with fast personal computers and good communication links between homes, companies and other users the use of the home as a production unit (for information) is likely to increase. This is discussed further in section 9.7. Firstly the typical use and flow of information is discussed.

9.3.2 Typical information flows and usage

The typical sizes of information objects and their use are relatively well-known in terms of the objects that are current in everyday life. However, in terms of multimedia communication these need evaluating in digital formats. At present, much information is still communicated in an analogue form and the use of digital encoding is restricted. The most common digital form of communication used in the home is probably the audio CD. This being the first mass production digital format used. There is also Teletext which is broadcast with television signals. However, there are soon to be digital radio and television channels which will increase the amount of digital information received by the home by a considerable amount. Further digitisation of information into the home is also likely with many newspapers and magazines planning electronic editions and user services.

To assess the impact of digital services the various media types need to be quantified in digital terms. To give some indication of the size of objects or the transmission rates of continuous information typical objects will be considered, i.e. a newspaper, a roll of film, an audio CD, and a television channel.

Newspaper. It is difficult to determine what is a typical newspaper. There are a number of different formats and styles used. Some use a higher proportion of images to text than others. For a typical example of a newspaper which uses mainly text it will be assumed that one issue of the newspaper will consist of 24 pages, in a broadsheet format, each consisting of about 20% image, 80% text. The majority of images would be monochrome although colour is often used on cover pages. This gives approximately 40 Kb of text per page and a similar amount of image data. This would be approximately 2 Mb of information for a single issue. Magazines would be much more data intensive since they usually consist of more image, more colour, and many more pages.

Roll of film. A roll of 35 mm film as used in many of the current home cameras is a typical format for images used in the home. A typical film consists of 24 images each with high definition quality and capable of being developed into a large positive image. So the 24 images will be considered to be typical large images of 250 × 200 mm. These would require a quality of at least 25 dots per mm to print adequately and use 24-bit colour resolution. This gives a digital form for each image of approximately 3.5 Mb per image or 84 Mb per film. The photo-CD standard allows many images to be stored on a standard CD-sized disc and this is currently used as the digital form of film equivalent to the 35 mm roll.

Audio CD. This has been in digital form since replacing the analogue formats in the late 1980s. A single CD is approximately 650 Mb of data. This has been a convenient form of packaging for home use and is now the dominant type of audio package used for home consumption. The stream rate of the audio output is approximately 2 Mbps with the included error correction information.

Television. A television channel consists of both an audio and video signal. The audio signal is relatively high quality and is commonly available in stereo providing two channels of audio. The video stream is between 25 and 30 fps each consisting of frames which would require approximately 1.5 Mb of digital information each. This produces about 40 Mbps of transmission required for digital television, until compression is considered. This will considerably reduce the data capacity require of any channel into the home, or alternatively allow more information to be received. The use of compression is thought to be inevitable for the widespread use of digital techniques in video distribution. The options for information provision seem to favour more channels rather than higher quality of image.

The other factor that is likely to be relevant is the rise of more integration of media types and the use of multimedia communication technologies in the home. This is the outcome of a number of converging technologies which are now finding applications in the home. It is also the result of a good deal of commercial research into products that can be used in the home, and information channels that can provide saleable alternatives to today's media and information sources.

9.4 APPLICATION OF MULTIMEDIA TO HOME INFORMATION

The information that is currently used in the home appears in numerous diverse formats and forms. The different delivery mechanisms deliver the type of information they are designed for and the differences are largely determined by historical convenience rather than current necessity. Current research has therefore focused on the different types of information and how they can be delivered to the home in a more integrated fashion. For example, there are currently a number of projects concerned with video-on-demand or the electronic newspaper. There are also specific projects looking into home-specific delivery systems. (These will be discussed further in section 9.5.) The other factor in the use of information in the home that will have an important bearing on the future is the degree of integration of the different media types into multimedia documents and the use of multimedia communication technologies to assist in its delivery into the home.

Currently the information used in any home is based around video, audio, image and text from different sources in different formats. A home will use television, newspapers, radio and other media to obtain information. These different sources all use different media to convey similar information. They can often contain similar information in different forms but are rarely linked. The use of a multimedia approach to information is currently not fully developed but can be seen to have been developing for a number of years.

Television stations already use two media types to convey information, a video stream and an audio stream. There is also a limited amount of text information that can be displayed on the screen at the same time. However, when the use of Teletext is considered the gap between current TV technology and a full multimedia approach becomes much narrower. Teletext allows the simultaneous display of text on-screen as subtitles, or multiple pages of related information to a specific source of information such as a particular programme. This extends the programme into more of a multimedia experience than is possible when relying on the real-time showing of video and audio streams since the Teletext can be broadcast for an indefinite time after the television programme is scheduled. This provides links between the programme and other resources which would require much more time to provide without the Teletext alternative.

The Teletext system is also a type of hypertext system where pages can be linked by use of facilities such as Fastext. In this system individual remote control buttons can be programmed to access related pages of text to allow a hyperlink to be formed, saving the user the effort of remembering page numbers as links between documents. Although Teletext has all the facility to do this, it has never been considered more than a free add-on to the television service although it does provide a large amount of standard information. The main drawback of the system is the poor quality of the graphics that can be displayed on screen. This caused by the design of the system and although other systems have been proposed the current Teletext seems likely to be around for a number of years yet. The lack of good graphics limits the amount of image information that can be displayed on Teletext. The best that can be done is a crude map with text overlay to show the weather. The system design does not allow a better display. If it did allow good images it would be nearer to a multimedia delivery system, but the bandwidth required to transmit the extra information would be greatly increased from current requirements.

Is it only a short step from a system that can deliver video, audio and text in this manner to a full multimedia system? To answer this depends on the user's perception of multimedia and the degree of its inter-linking and cross referencing. Most users would define multimedia to be *any* resources linked together that can be played back or viewed on demand as the user determines. This is different from the case of television where the system is essentially driven by the broadcaster. There is, however, room for both types of system. The devices needed to access multimedia would require more in-built intelligence to control the different software systems that are used to display the different types of information media and are likely to be more expensive to acquire than the much simpler television receivers. There are some features of multimedia that will be accessible from a television receiver but these will require the added intelligence necessary to be provided in an additional hardware device. This is discussed further in section 9.6.

The main point of contact between the user and multimedia systems for the next few years will be the PC or home computer. This already provides the type of software that can access information in all its varied forms and formats and can display it under user control on a single screen. Whether this is the way forward is debatable and stimulates many discussions. More discussion is in section 9.7. For the present there are two main schools of thought, those that advocate the use of PCs and other computers in the home and see this as the inevitable way forward, and there are those that work under the premise that the current home paradigm will not radically alter and that the use of the television set will play a major part in the future of the home and so work towards providing it with greater functionality via add-on devices and more flexibility of operation. Both of these camps are probably right!

With the trends as they are the use of multimedia is inevitable in the near future. The form it will take is difficult to predict. The evolutionary development from television to a linked set of resources using Teletext in addition to video/audio streams is likely to be extended as is the use of multimedia on PCs. Both of these have different applications at present. Most PCs are used by individual users and are designed to be single user devices. Televisions can be used in a communal setting, either by one or more members of a household. This difference in use is likely to be the main area for divergence between the use of multimedia for PCs and in other settings. For example, the use of linked documents is an individual process, reading text and viewing images are mostly done at different speeds by different people and they are difficult processes to use in groups. There are some situations when this can be done but they are not the main use of these media types which are more easily accessed individually.

Using multimedia in the home will then be a process of providing the information the users want to access in a form that they require in the different settings of the home which are commonly used. So educational material will be best provided for individual use by users on PCs. This allows for the provision of information in a hyper-linked form which can be accessed easily on any PC and other home computer. Leisure material provided for communal viewing will be in a format for use on television equipment and will have less hyper-linking as this is inappropriate in a communal setting. The integration of these approaches is discussed further in section 9.6. What is clear is that the provision of multimedia information for home consumption will require more capacity than is currently available in the majority of homes.

9.5 DELIVERY MECHANISMS

The current delivery of information into the home takes a number of routes, both paper-based (newspapers and magazines), via electrical signals (telephone) and electromagnetic waves (television and radio). To integrate these into a single channel is one of the aims of many researchers in this area. To do so requires that the delivery mechanism can cope with the capacity requirements of multimedia communication. As will have been seen from the earlier discussions in this book these vary widely for the different data types. Mainly text-based information is relatively low in capacity requirement whereas live video is very high. For home use the possible delivery of material for off-line use is an option. This allows data-intensive material to be down-loaded and used later. This can be done using much less data capacity (but more connection time) than where online information is accessed. The delivery of information in this form would require a different approach to its provision than a system designed for real-time use. Both, however, would require more capacity than is currently available in most homes.

The telephone is the main entry point for duplex communication at present, since broadcast media are essentially simplex channels. This has a very low data capacity in its current form. Even an upgrade to ISDN only gives an increase from 28 800 bps to 128 kbps which although a large increase cannot cope with full multimedia communication that may be required in the home. The solution is predicted to lie with the installation of cable and, particularly, optical fibre networks. This does, however, presume that the funding of such a scheme is sufficient to allow the provision of the services that the users require. At present, most cabling schemes concentrate on providing the basic services which are currently used by the home, i.e. a number of television channels and a basic telephone service. This should develop into a more multimedia-like service as the networks spread and the emphasis changes from infrastructure to content. There are, therefore two main alternatives for home delivery of multimedia communication services. The cable television network or an upgraded telephone system. Both could be delivered by optical fibre, but this is not necessarily going to be the case.

The technical issue is straightforward; optical fibre is the most suitable carrier for data-intensive applications such as multimedia communication and this would be the best solution for all homes. The possible providers of solutions are the cable television operators and the telecommunications providers. To enforce competition, the governments of some countries (such as the UK) have excluded the main telecommunications providers from offering this type of service until the cable operators are established. The cable operators need to invest primarily in infrastructure rather than content at the early stages of development of the networks and so the customer does not have the option of choice at present. The services are limited but the potential is there for later expansion. When the networks are nearer full operation the financial considerations will determine the services provided.

There are some interim solutions that are worth consideration at this stage. Firstly the use of ISDN does provide enough capacity for many applications and can be seen to be the transmission medium of choice for home-based multimedia in the near future. The limitations of a limited channel capacity could be overcome by using more channels and aggregating the capacity but the current cost of equipment to do this is prohibitive for home use. The other alternative is a new protocol developed for home-based systems. This is ADSL or Asymmetric Digital Subscriber Link. This uses the type of connection that is currently used for telephone traffic (within technical limits). The traffic on the link is divided between the two channels (into and out of the home) so that there is a larger capacity into the home than out of the home. The capacities are of the order of 2 Mbps into the home and 64 kbps out of the home. This has some good and some bad features. The positive aspect is that it provides a better capacity than any other method using current systems. The negative aspects are that the asymmetry militates against the home user who uses outward communication to do business or contact other users. This is not seen as a problem by many service providers since they require high-speed access to the home through which to sell their services. Those that are critical of this approach are generally those that have used home communication systems for two-way communication. It also limits the possible use of these systems for future home-based functions. This area is discussed further in section 9.7.

9.6 SYSTEMS INTEGRATION

One aspect of the integration of information into multimedia is the integration of the delivery systems into multimedia communication devices. The media types now used separately in the different information media that are used in the home can all be integrated into a single device, the PC. Text, image, audio and video can all be used, processed and displayed on a PC. This, in turn, allows the PC to act as television (with Teletext), radio, telephone, newspaper, magazine and photograph album. This degree of integration is generated by the all-purpose nature of the basic PC. Multimedia PCs can process and display all the different media types and allow the user to link between them creating multimedia presentations. At present these functions are all carried out by separate devices for specific purposes. The telephone is used for voice communication, the television for the receiving of broadcast television signals and Teletext, the stereo system to playback audio CDs etc. The home needs to have all these devices in order to receive and use information sold or provided in these forms.

The integration of all these devices into the PC raises a number of questions: Is it the best way forward in the home environment? Is an integrated solution the best solution? Is there merit in a distributed approach using separate devices? What home infrastructure is needed to make best use of these devices? Do interfaces exist to allow easy interaction?

Is it the best way forward in the home environment? The use of technology in the home has always been an interesting area for research. Much technology has not been designed to be used easily until recently. There have been many attempts to provide technology to make life easier in the home. Access to information and the exchange of information will form a major part of life in any information-based society. It is, therefore, logical that the ease of use and integration of devices to access and process information is a high priority. Integration allows use of information across the media types and the combining of different items into multimedia documents. With separate solutions this is not as easy and often not possible. In the home the need to save time and effort is as pressing as in business and elsewhere. Integration is, therefore, a useful tool of the home information user provided the options are fully considered and the home environment does not suffer as a result of the introduction of a new paradigm.

Is an integrated solution the best solution? The use of an integrated multimedia package allows the user to have a standardised interface and customise the software and hardware to suit a particular working style or function. When the equipment is in a home the integration is less important than the actual use and this tends to suggest that users will be more likely to use separate pieces of equipment than an integrated package. Consider what would happen if the home computer were being used to access an educational package and the rest of the household wished to watch a film! This may indicate that a number of separate integrated multimedia devices would be needed but in some situations a cut down version (such as a television receiver) may be more cost effective! So the technology to use needs to be the appropriate technology for the application that is being used for the information that is accessed.

Is there merit in a distributed approach using separate devices? The use of separate devices will give the home a distributed information system with different types of information being delivered into different parts of the system. There may not be any connection between the different devices. If there is, it may be difficult to transfer information between different formats. The integrated solution aids this process, but it is not the only way to exchange information. The use of separate devices can include data interchange but is usually more difficult.

What home infrastructure is needed to make best use of these devices? To enable more efficient use of information in the home there would need to be some way of connecting devices together. The integrated solution does not require any special thought on this matter but with separate devices from diverse sources the problems of standards and compatibility arise. To connect diverse devices in a home the information can be exchanged in a number of diverse formats or a structural solution can be used, such as a home network. These are, however, rare and still the subject of research.

Do interfaces exist to allow easy interaction? The ease of use of any home multimedia information device will affect the use to which it is put. The standardisation possible on a PC interface and ease with which users can tailor interfaces is one of the main positive points of the integrated solution. Individual devices also need a simple, helpful interface to allow users to access information as simply as possible. The various technologically-based information sources have all attempted to provide users with suitable interfaces over the years: Teletext systems have developed the Fastext facility, there have been many attempts to create a good interface to the video recorder such as Videoplus and Programme Delivery Control, and there have been changes to the telephone to make it more user-friendly. New devices will need to build on this experience to give users a logical interface to the complex web of information sources that are available.

The integration of different information handling capacity into a single device is common and used frequently in multimedia PCs but this does not mean that the home based solution for multimedia information is inevitably going to be an integrated one.

The appropriateness of the technology is more important than the drive towards integration at all costs. The user should drive the technology, not the reverse!

9.7 NETWORK POSSIBILITIES AND PROBLEMS

The application of multimedia communication technology to the home environment does more than automate the current method of accessing and using information. The additional power of high-speed connections to multimedia communication networks and the ability to process large amounts of information allows new possibilities for the home and the extension of some uses for the home into a more flexible way of operating. For example, teleworking has been possible for a number of years using telephone lines and computers but higher-speed networks allow a greater range of work to be carried out from home than was previously the case. There will, of course, be problems with the installation of any new technology into the home and some of these have been the subject of some speculation.

9.7.1 Re-shaping the home environment

Some homes are used solely for domestic and leisure purposes and some also act as a work place. This use of the home has a long history. The amount of work taking place in the home that has changed in cycles with the different cycles of human organisation. The early large scale organisations of society were based upon agriculture and most workers were needed in fields to grow crops and tend animals etc. Mostly this was done within a short distance of the home and some of the additional crafts were carried out at home (or in an adjoining workshop). For example, the smith would live in the smithy and each village would have a smith. When agriculture became more automated and industry began to grow there were moves away from the home into factories for many workers. There were still some exceptions. For example, chain and nail making was often carried out from home workshops in the 19th and early 20th centuries. As society changed and became even more mechanised and automated the need for workers in factories diminished and the work that was required became more based on information than was previously the case. This change is still taking place in the late 20th century. The society of the western world is becoming much more an "Information society" with workers processing information and providing other workers with information while the manufacture of goods is increasingly automated until it requires little intervention from workers. This significantly alters the way work is done and where it can be done. Much more of this type of work can be done in a home environment and much more can be operated from home even if it requires a visit to an external organisation to perform certain tasks. For example, the operation of directory enquiry system for a telephone operator does not require attendance at a particular place, the operator can work at home and the calls can be re-routed using the communication network. Also, an organisation of service personnel such as gas appliance service engineers need not work from a depot but calls can be re-routed from a central number to individual homes or mobile telephone numbers.

The increase in work in information-related areas has allowed more workers to work from home, as teleworkers, as more work is based on information rather than on physical activity. This increase has not been a straightforward change from one system of work to another but has been much slower than many observers have expected. It seems that workers are often reluctant to work from home and this causes them to keep to the more traditional (20th century) paradigm of working in offices or factories. The home is seen as a leisure domain rather than a work domain and the two are kept separate. However, the introduction of multimedia communication networks allows a much better quality of information to be delivered to the home and the pressure to change to teleworking will increase. This will affect the home and the work environment in many ways, some of which may not be foreseen when a large mass of users change to this way of working.

The effect on the home may be greater than is thought to be the case by those that push for the uptake of teleworking. A number of studies have found that teleworkers find it difficult to separate the home and work environment when engaged on telework and a number of guides to how to organise the work sector of the home environment have been published, mostly these concentrate on reinforcing the work methods used in offices into a home environment and separating the two domains as far as possible. This may be a sensible alternative at present but when masses of workers start to use teleworking this may not be the most appropriate form for self-organisation. The tendency of homes to become workplaces will make home organisation change but the scale and nature of such a change is unpredictable at present.

The home is an individual entity. No two homes are the same and the differences are usually in details that are important to the users. The changes that will be possible in home environments with the introduction of multimedia communication technology could range from insignificant details to fundamental shift in the way people live. What is probable is that any change will occur gradually and new technology will be taken up as it becomes useful and the living practices that take place will adapt to suit the environment needed by the occupants. Many observers have said that large scale change is inevitable and that the new paradigm is only a consequence of this technological upgrade. This is not necessarily the case, but there are strong arguments in its favour. Further discussion of the societal implications is in section 9.7.3 and 9.7.4.

There are other implications of the technology which may require answers in such a home based on technology. If the occupants are engaged in telework what will be the effect of a work and leisure environment based around the same or similar machines? Will the two different environments interact or interfere with each other? Do users at home want this type of facility which will keep them by a screen all day? How will these systems be shared between users who want to interact with the technology? Who decides the use when two users have different immediate needs? For example, a typical situation may be when one user needs to use a system for work-related activity and another needs to access educational material. These questions are difficult to answer without clear examples on which to base further conjecture.

9.7.2 New services

The use of multimedia communication in the home gives rise to both new services and new methods of doing home-based activities. These could lead to totally new home paradigms and it may be that the future is radically different from the present.

New services. The type of new service that can be provided is inevitably based on current paradigms and practices. Currently available services that require high bandwidth, such as videophone calls and video conferencing are obvious candidates for use in the home where people communicate with each other when separated by some distance. Allied to video conferencing will be the use of image and file sharing to exchange image and text information with the video call. Although these applications are currently used in business environments their use in the home is just as relevant and in some cases more urgent or useful. Other services that may be provided can come from the whole range of media types. Electronic newspapers are an area of current research and development with some prototype examples being available. There are many ways to distribute information into homes, some of the current schemes may continue, some new ones will be developed but the trend towards more information being provided via electronic digital channels is set to continue. One of the main areas being investigated by service providers is that of video-on-demand. This is where video films or programmes are stored on a server and the user dials in to download a video to their television via a telephone connection and some specialised hardware such as a set-top box. The particular advantages of this are that users will be able to relate easily to the technology since it is a direct evolutionary step from television remote control and satellite decoders to the new set-top boxes. There are various plans to expand the services provided through the box to encompass home shopping, home banking and other familiar services. As usual the predominant application is in the entertainment field as this is seen to offer the greatest return on the investment needed to start such a network of interactive home-based communication. The other services are likely to follow in the wake of the main entertainment offerings.

New methods. The possibility of performing home-based activities differently is also of interest. Many activities in the home require a lot of effort to complete them successfully or are hampered by the lack of information. The regular contact between people has been aided by the use of the telephone but there are times when audio information is not sufficient to convey the level of information required. To some extent fax has helped in this but it has not been a major application in the home as it has in the office. The paper-based mail service is still more often used for home-based text transfer such as letters. The advantage of multimedia communication networks will be in the range of media types that they can convey opening up many possibilities for users to communicate. For example, a family may be temporarily separated by distance and one member may be away from home. A multimedia network connection would enable them to take part in any important aspects of family life from a distance. This allows a home to become stretched to include remote members as if they were in the physical space called home. This is the virtual home concept which has been the subject of some discussion and research (see Sloane, 1994). There are also possibilities for pursuing new leisure activities related to the networks themselves. One of the findings of various researchers who looked at the French experience with the installation of Minitel terminals in the home was that the users soon used the services that allowed communication between people such as chat lines when the initial installation drive was for access to database information. This leads to the conclusion that users will use such services as far as is possible to follow their own interests and if these are not provided for then the service may not succeed.

The home is set to change with the introduction of this type of new communication technology, how much and how soon cannot be predicted without some evidence from actual installations. The total effect could be enormous with the home becoming a new and different entity from what it is at present. Alternatively, there may be little change which could stem from an in-built resistance to change. However, even small changes in homes can radically affect the face of society at large. The shift to new work and leisure patterns may not be a great effect to the individual but could be significant to the whole of society.

9.7.3 The virtual community

A number of writers have looked at the possible changes to society that could result from the greater emphasis on information that is currently taking place. The information society is a concept that is now recognised by governments as being a way of society organising itself in the future. Some of these writings have been mainly technologically driven and some have been more biased to the personal aspects of such a change. There are, however, two main camps. Those that see the technology as being the only way forward and those that warn against such profound change. This section looks at some of the more optimistic predictions, the others will be discussed in section 9.7.4.

An early view of the future was put forward in 1980 by Toffler (see reference at the end of chapter 5). This work predicted the rise of the producer-consumer in society when information became the commodity that predominated in the business world. The premise was that users would retreat to a home-based economy where they would both consume and produce information to provide an income for the household. This was a prediction of the "Third Wave" which equates to the information society. This scenario is still many years in the making but the role of multimedia communication is a crucial one in any realisation of such a prediction. The type of society envisaged where users are information processors is dependent on a widespread communications infrastructure, such as the proposed information superhighway, to succeed. Whether it is what is wanted or what will actually happen is, as yet, unclear.

A study based more on present-day activities is contained in the book by Howard Rheingold referenced at the end of this chapter. The idea that computer networks can create virtual communities is clearly one that has been fulfilled to some extent and the users of these networks can often be said to be in some distinct community when on-line. These communities are different from those based in the real world. The basic structures and ideas in these virtual communities are based around simple technologies and may change when full multimedia communication technology replaces the technology used to create the communities studied. The changes wrought on existing communities may lead to new communities forming and these may be based around similar ideas to those that led to the first virtual communities that were formed. Often, what is neglected in many studies is the realisation that communities in real-life still exist and that their formation and interaction are still important to every-day life. A virtual existence is no substitute for the real thing.

The impact of the technology upon individual may be small. An individual user may change their personal schedule from watching television for an hour or two to using a computer communication network for the same time. This may not be a highly significant change, but if a few hundred thousand users do the same the whole structure of this new virtual community (or communities) comes into existence from nowhere. These communities, based on communication networks, do not necessarily meet in real situations and all the information that flows within the community is based on simple computer communication networks. With the use of multimedia communication networks these communities may have more impact and create a more realistic virtual existence which may not be to the advantage of the participants. The implications for the rest of society may be profound with the use of communication in virtual environments being preferred to real communication in face-to-face situations. This may seem extreme at present but the change towards more isolation of the individual is taking place as society becomes more dependent on electronic and other media to present views of society that are outside most people's everyday experience. For example, perceptions of violence in society are often much higher than the actual risk to the individual. This can be for many reasons but it tends to increase isolation among the elderly and more vulnerable in society. The addition of communication networks that would enable a home-based existence may reinforce such perceptions as being real.

9.7.4 The home fortress

This view that the home may become more like a fortress is one of the more disturbing features of a move towards greater use of multimedia communication networks. If the world moves in this direction with more homes being linked together; do its inhabitants gain from the connections or lose by them? Multimedia connections (like television has done) will cause homes to be more isolated as leisure increasingly becomes an activity that is carried out at home. Activities that require interaction between people will be able to be carried out at a distance and this will tend to increase the isolation of homes and a fortress metaphor will be more applicable. However, people do not necessarily want to use communication systems for all contact with others as the quality of contact is less than in a real situation. The gains from such systems must not be ignored, many users are unable to interact effectively with a wider community because of isolation or disability and multimedia communication systems would enable these individuals to play a fuller part in society than is currently possible without the technology. Even for individuals that are not disadvantaged there are benefits to using such technology, with better information than is currently available via technological channels and more choice in the information accessed or used the individual may find significant enhancements in some areas of home life.

One final point that is often debated is the proposition that such a system may encourage greater division in society between the information "haves" and the information "have-nots". There are many areas of the world where the currently available technology that is commonplace in the western economies is not available on the same scale. This could have one of two effects. These countries will either spend a long time trying to catch up with the technology (if they start), or they may be able to leapfrog the technical position of the western countries and install up-to-date technology without going through the same phases as those countries that have used the current technologies to the full. This may make the latecomers more advanced in a shorter space of time than those that adopted that last phase of technology and are reluctant to change or upgrade to new systems. This effect has been noticed within Europe where some countries implemented standards for EDI when they became international and became better connected because of their adoption than early users who were left with too many standards to interconnect!

The use of multimedia communication technology may have unpredictable effects upon homes and society, some of these have been discussed, some are not yet apparent and will emerge as the technology is implemented in homes. It is clear that any introduction of technology will change the nature of the home. There are examples of this happening in the 20th century. These may not be serious and may even be beneficial to both the home and society. What is necessary is a clear understanding of the problems and a willingness to acknowledge them by the technologists. One further reference that is worth mentioning is the book by Slouka referenced at the end of this chapter. This takes a more human approach to the technology of communication and is an interesting discussion of some of the possible pitfalls of this technology.

9.8 SUMMARY

This chapter has mainly concentrated on the non-technical aspects of multimedia communication technology in the home. While this is not yet a widespread technological option it will become more available as the technological spread into business and education slows down and the vendors and service providers look to the home as the next main market. The options that may be available will be mainly based around leisure and entertainment at first and the successful implementation of any communication system will depend on the infrastructure that is in place for home users. Much will change as the technology changes and predictions about what may or may not happen are not really possible without direct experience of the effects of the technology.

FURTHER READING

Much of the current thinking on the effects of communication and information technologies on the home is carried out by members of the IFIP Working Group 9.3. One of the best condensations of their deliberations is contained in

K. Bjerg and K. Borreby (eds.), *Home-Oriented Informatics, Telematics and Automation, Proceedings of a cross-disciplinary working conference*, University of Copenhagen, June 27th- July 1st, 1994

Two opposing discussions of the virtual worlds created by multimedia communication technology are discussed in the following two books.

H. Rheingold, *The Virtual Community*, Secker and Warburg, 1994, ISBN 0-436-41214-4

M. Slouka, *War of the worlds: the assault on reality*, Abacus, 1996, ISBN 0-349-10785-8

Specific references

A. Sloane, Homelink: Technical considerations and proposals, in Bjerg and Borreby (op. cit.), 1994.

EXERCISES

1. During a typical day note the different types of information that arrive in your home through all the different channels. Assess the relative size of the information input through each separate channel or information source.

2. Calculate the size of channel needed to obtain all the information used through a single channel. Assume appropriate compression levels are used.

3. Calculate the information storage requirements for your home. Include all current sources of information such as books, newspapers, magazines, photographs, audio CDs, videos etc.

4. Design a home network for a home consisting of a group of two adults (who both have employment) and two children (who are both in school).

MULTIMEDIA COMMUNICATION AND THE INFORMATION SOCIETY

Summary: This concluding chapter shows how the various application areas interact together - linking digital libraries to teleteaching, teleworking and home-based users. A look at trends in tools and systems and a final look at various societal implications of digital multimedia communication.

10.1 SUMMARY

This final chapter is designed to bring together the various areas of interest developed in the book so far. The interactions between business, education, the home and information are becoming more complex as new systems are invented and used. This discussion of the various interactions that are inevitable between the different areas of study and application allows them to be discussed more fully. Without an analysis of the interactions the different applications act as islands, cut off from the rest of society and functioning without reference to important sources and outlets for information. The overall idea that is behind this push for integration and the spread of communication in the world is the evolution into what is now called the information society. This idea will also be developed further in the next section.

The various application areas covered in Chapters 5, 6, 7 and 9 are only some of the areas that can benefit from the introduction of multimedia communication technology. The areas of healthcare, manufacturing, publishing and many others have been left out of this book but are all equally valid areas for looking at the impact of these new technologies and the effect they will have on the operation of the areas in question. Some of these areas will also be used in this final chapter. What is also clear from these earlier chapters is that there is an effect on society from the introduction of the technology that will need to be assessed before its widespread use unless the problems that have occurred with previous technological introductions are to be avoided. This presupposes that the effects are able to be quantified or even predicted. In many instances there are unpredictable effects, both good and bad that occur with the wide scale introduction of technology. For example, the introduction of road and air travel has allowed more people to live further from their employment than would have been possible in the late 19th century, this in turn causes more pollution and accidents on the roads but it also allows holidays to be spent in sunny and exotic locations with no more time expended on travel than a cross-country trip to the coast. Sometimes it is quicker to go to a sunny country by air than to travel by road to the nearest seaside resort! Both of these effects are the result of the introduction of technology on a wide scale and a number of other factors that interact with the technology and society's use of it.

The rest of this chapter will not be able to predict events in the future and it is not attempting to do so. What it does cover is the possible effects of individual technologies as they are now and what improvements will mean to users. The societal effects will be discussed but mainly as a consequence of what is currently known about the type of technology under discussion. This is essentially a fine line between the effects of technology at present and the area inhabited by futurologists and astrologers! So far the technology described has all been feasible and in many cases in operation. Some of this technology is not intended to be used in some of the situations described in the book but they may benefit from its use. For example, many technologists consider business more important than the home and reserve the faster speed data connections for business use. However, many of the data-intensive applications are very prevalent in the home and these views can be a little short-sighted, especially when the actual use of the technology in business is considered and the data rates calculated as they are often far short of the typical home use, especially in the video area.

10.2 THE INFORMATION SOCIETY IDEA

What links together the technologies included in this book and the various applications is the overriding consideration that society is changing from one based on a system where most workers produce goods for sale to one where most of the workforce is engaged in work that uses information or provides some value-added service. This is usually termed the information society and has been briefly discussed in Chapter 9. The information society has many different aspects and much of its evolution depends on the incremental use of technologies that improve communication between people, between computers and between computers and people. Most of the technological problems and features have been discussed at length in the earlier chapters, and the applications in some areas were the subject of others. The integration of these technologies and application areas is where the information society can be seen to be most evident at present, and this presents a model for the future use of multimedia communication technology.

The information society is a broad definition of how society will function. It can contain many different implementations. Some observers see it as being a complete transformation, some see it as being a small change from what is currently done. The probable outcome of any change is that small changes in technological options will be introduced and taken up by users and this will then have some effect on the way society functions, but it is not until a critical mass of users forms, in any area, that the profound effects become noticeable. For example, the invention and subsequent introduction of the telephone in the 19th century had little initial impact and most business was conducted face-to-face and on paper, but when the majority of business users and homes became connected, the actual processes of business changed to adapt to the technology. Now there are specific jobs related to the telephone and some users work full-time answering and making calls. This business would have been done very differently without the telephone.

There are many other examples of technology changing the method of work and leisure. Consider the changes that have been wrought with the widespread use of television, fax machines, mobile telephones, etc. Some technologies have a more profound impact on society than is first imagined. Equally, some have less! Those that do have an impact are difficult to predict when they are new and largely untried and this is mainly the case with much of the multimedia communication technology now available. It could be that a great number of users take up the technology and use it to replace current methods and practices. Alternatively, it may remain a technology that is used by relatively few people for particular tasks that require the use of the technology and that cannot be done in any other way. Some of the factors that affect this take-up are: the technology itself, the ease of use, the sale, pricing and marketing of the products, the interfaces used, the actual uses and convenience of use of any associated equipment and many more. Each of these points is worth further expansion.

The technology itself. To have any real prospect of take-up a technology must be of some real use to the people that are going to use it. The telephone is a good case of this. It saves a lot of time writing and answering letters or making journeys to face-to-face meetings, so it is deemed an essential tool of both business and the home. It is this level of usefulness that makes a technological product become an indispensable tool. Many products have been proposed to be as useful as this but few have actually achieved such a level of use as the telephone.

Ease of use. To use technology can often be difficult. Not only do the concepts behind the products need to be understood but the actual operation of the different types of devices also need some comprehension before they can be used effectively. Many devices become easier to use with knowledge of their operation, but unless the operation is standardised with other devices difficulties can still remain when devices are changed or functions added. Standardisation plays a big part in removing the uncertainty from the operation of many new technologies. For example, the standardised interface to the telephone allows users to perform the basic telephone function easily, i.e. dialling a number, and although other functions may be available the simple use of the device is not constrained by the other features. If this is compared with a computer where some understanding is required to do anything useful then the features of easy-to-use technology become clearer. Standardisation is only one part of the required package. There also needs to be clear design features and simple user interfaces to make it all work.

The sale, pricing and marketing of the products. The actual take-up of any technology will depend on other factors outside the technology itself. A product may be wonderful but if it cannot be sold because the price is too high, no one knows its availability or it is difficult to obtain then the technology will not have a great impact. A good example of this problem has been the take-up of ISDN channels in different countries throughout Europe. The different rates for installation and usage of ISDN has given the take-up of ISDN a different pattern in the various countries that offer the service. Linked with this is the cost of equipment and terminal adapters. The pricing and marketing of this equipment and the associated digital services has led to a curious position for ISDN which could have replaced the ordinary telephone line by 1995 but has had little real impact except in a few countries.

Interfaces used. The interfaces are as important as any other factor for the take-up of technology. The design of an interface can greatly affect sales and subsequent take-up, although there are cases where the interface is of secondary consideration, for example in the video cassette recorder. Mostly a good interface will help with the subsequent take-up of the technology and an indifferent interface will reduce the effect the technology has on normal day-to-day existence. Again, the telephone is a useful example since it has one of the simplest interfaces for the functions it performs.

Actual use. The actual use of technology, when it reaches a wide audience, is not always what is envisaged when it is first made or proposed. Some technologies are widely applicable to different functions and it can be that a secondary function becomes more useful than the primary objective of the design. The telephone was initially envisaged as a broadcast device but soon became a point-to-point communication device.

Convenience of use of any associated equipment. Alongside the interface is the convenience factor of any technology. A piece of equipment must be usable in the required situations or it is not as applicable as it may be. For example, early mobile telephones did not have a great take-up since they were heavy and difficult to carry around when they became smaller and lighter they soon became used by those with the need for mobile communications. Price and marketing also played a part that cannot be ignored.

The information society will be built on some technologies. Exactly what they will be is hard to predict at present, but some of the current technologies that are new could have some particular effects on the way of business and the way of life of many people. For example, multimedia communication technology with video conferencing has been hailed as the next step in communication technology for many years. It was not until ISDN-based systems became widely available that the technology became both cheaply available and widely available enough to consider it as having any impact. It will, however, require more sales and promotion before it has a great effect. The effect of this one technology may or may not be great. Other technologies may be more profound. To build an information society will require many different technologies to be used by a majority of its citizens, in the manner of the telephone at present. It is only when the core information society technologies are in place that it can be said to have happened. These technologies may be open to debate but some will be more obvious than others and the ability to exchange all types of media and information will be fundamental to the start of such a societal shift.

As indicated in the previous chapter there are many factors which are required to change society in the way indicated to encourage a move towards a higher information base. There are a number of ways in which this information society could come about. Different technologies could shape the information society in different ways and the shift could find resistance from current vested interests. However, the general consensus among many different players is that the shift will occur. The amount of change is not predictable and neither is the timescale for change. This latter factor is likely to be different in different countries as there is already evidence of different perspectives and initiatives being carried forward in different parts of Europe, North America and the Pacific Rim countries. These differences in implementation and emphasis may make the information society more problematic than is yet known. Any disparities between different sections of the community are likely to be exacerbated by a move to an information-based paradigm.

To work and live in an information society will require many of the applications discussed in the earlier chapters of this book to be in place and widely available to enable people to work and live in a manner that is more person-centred than currently is the case. The variety of applications needed will inevitably include much more than has been covered and some that are not yet invented, however, the basic building blocks are already available. The need to work with information is increasing, the use of more automation in manufacturing is also increasing. Together, these two factors build an information society. The organisation of that society is the next crucial factor in the acceptability of the idea among the general population. There are instances where the use of information society tools and methods are available at present but do not find favour with a large number of users. These are often due to cost factors but sometimes may be due to other, more subtle, influences. For example, a typical information society application is teleworking. Many people now do some part or all of their work by teleworking. There are, however, significant factors that prevent a wider application of these methods. Many office workers feel isolated in a telework situation where communication is less frequent between themselves and the rest of the workforce. This is often seen as a barrier, both to promotion and to the informal communication channels of the workplace. On the other side of the workplace there are managers who feel that teleworkers are not working as hard because they can't be seen! There are many jobs that could be done effectively by telework and this is often prevented by either management or the affected worker. This effect may be one of those that prevents the effective spread of the information society in some countries. Others may not be organised in the same way and may take a more helpful stance. There are also technological barriers to the spread of the information society. Companies are often unwilling to invest in developments that are not of direct cash benefit to them and a shift to teleworking may be in this category. The use of multimedia communication technology will enhance the quality of information available to users in an information society. Current technology that is in place does not have sufficient capacity to enable a significant shift to telework and encourage a greater use of communication networks. When the networks are significantly upgraded the situation will change. The other factors will, however, still be relevant. The ability to telework or process information from remote sites does not automatically make users want to use the technology. It is similar to the observation that the widespread use of the telephone has not removed the perceived need for face-to-face meetings between people.

10.3 LINKS BETWEEN DIFFERENT APPLICATIONS

The different applications covered in this book may all play a part in the information society and to make information useful to individual workers the various application areas will interact. For example, the home-based teleworker may need access to library facilities and the networked education establishment will need home communication facilities to make the application function effectively. Many of these interactions are relatively straightforward but they are worth further consideration to highlight areas in common. The various interactions of the four application areas of the home, business, education and libraries will be covered, others are left for the reader to fill in.

Education-business. The interaction between education provision and the work environment can be modelled on the current practice of many educational establishments and companies where the provision of training, research and consultancy services can be provided by an educational establishment and different companies can use services for agreed fees when required. What is possible with greater communication capability between educational establishments and workplaces is a more responsive system of fulfilling the needs of the business partner and at the same time allowing commercial information to be used more readily in the education process. Multimedia communication technology with appropriate applications can allow desktop, on-the-job, just-in-time training. It can also have a similar effect on the educational establishment where more up-to-date commercial information can be used from appropriate commercial sources. If other interactions are also considered the workplace user could be a home-based teleworker and the educational provider could also be based at home. The capacity restrictions that currently discourage all but the most simple form of co-operation are reduced and the ability of users to collaborate on projects of this type are enhanced.

Business-home. The interaction between work and home has partly been covered in the discussion of telework, but this is not the only business-home interaction. Business often relies on advertising to sell products and if the nature of the business requires selling to the general public a home-based presence is required. this is currently gained via television and newspaper advertising but if information is allowed into the home on a more selective basis then the nature of the advertisement will probably change. The use of interactive channels between homes and businesses will also change the nature of the communication needs of these areas. For example, online catalogues will benefit some users who may want to browse goods at home and this allows for inclusive advertising within the same area of display as the chosen goods and in other ways. A current example of this is the use of advertising in commercial Teletext channels. Here the advertising page is displayed between the information pages, or on the information page, making it difficult to avoid by the viewer.

Another area of interaction is in the provision of information goods. That is goods and services that can be exchanged in electronic form. These can be provided to home-based users via multimedia communication channels and payment can be made electronically. This type of service can be provided to homes from any computer connected to the network. This type of transaction is currently in its infancy and some reservations need to be addressed before a widespread acceptance is possible, but the possibility is real and likely to be more prevalent in the near future. This interaction between homes and business will lead to some changes in the way goods are bought and sold and a widespread use of such systems could have a profound effect on society having some positive and some negative effects. For example, more selling via networks will lead to a reduction in the need for large scale retail outlets for information-based goods and this leads to a scaling down of the distribution infrastructure required for the goods. This will reduce employment in this area but may lead to increased employment in the after-sales service of the goods, i.e. technical support and training.

Library-business. The interaction between libraries or large-scale information resources and businesses are not so obvious, but most businesses rely on large quantities of information to provide the goods and services that they offer. These information sources vary from telephone books which provide telephone numbers of clients, customers and suppliers or prospective customers, to information databases of different sorts providing useful business information to allow more effective production or administration. With these resources available online the business will reduce the time to market of the goods or services that they provide. The range of information will be less restricted than is often available to a business since the use of online resources can allow the tailoring of particular queries to the task in hand rather than having to purchase the whole database or directory. Many different sources of online information exist at present which enable businesses to function better, with a complete online resource the need for paper versions will be reduced as will the storage requirements for libraries of information on-site or near to the user. This in turn frees up the user to work remotely from home, or elsewhere using mobile communication technology.

Home-education. The interaction between home and the world of education is more than just the provision of distance education. There are many times when education is required which are not provided by current processes or structures. Although distance education is one aspect of the information society paradigm there is also just-in-time education, educational leisure activity and other educational activity related to work activities done from home. The use of education has traditionally been restricted to particular places where schools, colleges and universities are located. This is less relevant now as multimedia communication networks allow personal interactions to take place across the world with educators and other experts. Modern distance education has used a number of different media to provide a service to students. These can now be combined together in multimedia communication channels to provide an online service to users. The home, however, is a place which requires more than simple educational packages. The use of educational material may not be for the originally intended purpose and may be part of a wider interest that is not primarily educational. The other activities in the home may require educational input of a different type to that of the distance learner with information being provided by educational sources. This information could be seen as being of an independent nature from a trusted third party. For example, a typical activity in some homes is the decision on how to vote in elections, this may require information about the effects of policies put forward by different parties and these could have expert commentary provided by academic sources as a source of reference. (This may be a bad example of impartiality since few people are truly apolitical.)

Library-home. The use of libraries has long been associated with leisure activities with many public, and some private, libraries existing mainly to service the requirements of their users for leisure material such as novels, biographies and leisure-related non-fiction. Many sources of information are usable over electronic communication channels and this would allow a more rapid access to the information by the user with a more focused approach to its acquisition and use. i.e. users could download the information wanted in a usable form rather than borrow several books each of which may contain a single chapter of interest. Some information may also reside in homes where users have a particular interest and this could then be made available to others via the networks. This would provide a distributed information resource for users to browse although there are some security and privacy requirements to be addressed. So the nature of libraries and information resources could be changed in an information society, however, it seems unlikely at present that a large scale implementation of this type of information will take place soon. There may come a time when electronic information becomes much more cost-effective and the channels for delivery and use become more common, but as yet this seems a number of years away. To many people the main problem is one of useability. A book is easy to use. It can be taken into the garden, or to the beach, on a sunny day and it can provide a simple reading experience. Any electronic form of text has a long way to go to emulate such ease of use and versatility. In other situations electronic forms may be more useful. For example, instead of storing back copies of magazines in a home library a user could use the electronic version when and how they require. So if the user wanted to find out about new computers or some type of peripheral equipment they could query the electronic magazine for a suitable comparison rather than wading through a number of back issues.

Library-education. The link between libraries and education is long and historical. Many of the first libraries were connected to seats of learning, such as monasteries and the modern universities and schools have been based on ideas from those times. Now, any educational establishment needs access to a library of resources. These resources are increasingly in forms other than books, although books still form a large part of an educational library. At different stages of education there are different requirements and some of these are better served by non-book materials. However, it would appear that there is still a large market for the outpourings of academics in book form which is of great comfort to the author of this book! Mostly the electronic form of libraries is a welcome change in much academic work, especially at universities where small pieces of information are commonly needed from diverse sources. The use of networks allows this to be an easier way of working than with traditional sources. Academics can attach to the library of resources via a desktop computer and use the library database to search for references that are relevant to their current work or other projects. This does more than reduce the need to walk to the library, it allows a more instant response to information need and this gives the electronic library user a clear advantage over the paper-based user. The computerisation of library catalogues has provided some initial aid to the user but the full scale digitisation will provide more benefits.

In an information society it is easy to envisage a user working from home using a digital library to find out information for a current project and taking time to catch up with current techniques and practices via an online course from a university or other education provider. This scenario assumes a number of things about the implementation of multimedia communication technology. Firstly, that enough capacity will be available at the user's computer to enable it to function effectively with information available form the network which could be in any form from text to video. Secondly, that the organisational structures required to work effectively have been designed from home working and thirdly, that the rest of the structures are in place, e.g. digital libraries with a workable reference system. It is clear that the information society will take a lot of organisation if some of the more ambitious ideas are to be realised. What will generally happen is that parts of the technology will be applied in some areas first and then eventually filter out into other areas. The order of the implementation of technology into these other areas is indeterminate at present and will largely be decided by political or economic forces as the research in different areas is funded by commercial concerns or government bodies each with their own priorities.

10.4 EXAMINATION OF COMMON THEMES

From reading this book it should become clear that there are three main themes that have been applied to the various applications in the different chapters and that arose from the early discussion of information. These are the increase in the availability of digital information, the organisation of digital information and the implementation of multimedia communication networks. These three themes will be discussed in turn in the following three sections.

10.4.1 Digital information

The change in the way information is now presented on computer to the way it has traditionally been presented on paper and in other analogue media is connected directly to the ability to digitise information in different forms. This has been covered extensively in this book and is the core idea behind any of the modern forms of communication. Information can be captured, stored, transmitted and received in digital forms where the quality of the information can be determined and the user can control the information that they require. This is mainly due to the increasing speed and power of modern computers, especially desktop computers where many megabytes of information can be stored and used for the various purposes outlined in previous chapters. Without the ability to handle digital information in a convenient way many of the current systems would not exist. The digitisation of information has one important effect, that is that the different forms and formats of information can all be handled by the same devices and communication channels. Video, image, audio and text all become bits. Bits can be stored and transmitted by digital devices. There are some extra problems which need to be considered for the synchronisation of different media streams but for the most part the different media all become the same at this level. While the information is all in the same format the amount of data needed varies widely between the different media types. To use video requires a lot of bits. Text requires very few so there are different considerations for the transmission and storage of different types of information which are determined by the use of the information and its original format. Finally, the storage and transmission of digital information has given rise to a number of data compression schemes that can reduce the redundancy in information and make it easier and quicker to both transmit and store. These schemes allow much better information throughput than would be possible without them.

10.4.2 Organisation of information

One of the positive points of the traditional means of storing information has been the schemes and systems enabling access to it. These have grown up over the years. The information available has grown and the systems have changed to incorporate new ideas to permit access to the information. However, the use of digital technology to store, retrieve and transmit information requires new structures and systems designed to be used with the technology available. Unfortunately, as is often the case in the computer age the technology is ahead of the systems for organisation and there are a number of instances where information is available but difficult to access because of the shortcomings of the systems used to access it. A typical case would be the World Wide Web which has grown so fast. The access to information is via uniform resource locators which do not contain any inherent classification but show where information is stored. It's a bit like having a library with no catalogue and all the books are stored in shelves related to the publisher! Systems need to be designed to enable ordinary users to have easy access to digital information that is now available over networks and to be able to be confident that the information they have obtained is the best available for their needs. Libraries and other information providers have developed systems that work for paper-based and other media, the user needs systems for computer-based information. There is a considerable amount of information available in different forms from ftp archives, the WWW, particular databases and in newsgroups and discussion lists but little of it is easily accessible in the traditional sense. It may be available easily but the user needs too much knowledge to gain access to say that it is accessible. These problems will be solved, but the solution may be a little too late for some of the archives that are currently in operation as their re-organisation may take considerable effort.

10.4.3 Multimedia communication networks

Finally, the other main theme that has developed through the book has been the development and installation of multimedia communication networks. These are systems which have developed to carry large quantities of data and are the modern systems for communication between businesses, homes, education and other users. The current networks available do not have enough capacity to enable a useful communication of multimedia objects between users and information sources. This will not be the case as new technology becomes more widely available and multimedia communication becomes a more practical proposition. Some small networks, and some limited connections do currently exist that allow mass data exchange between users. These are expensive and require expensive equipment to connect users together. Some intermediate speed technologies such as ISDN have been around for a number of years but have not made a great impact for various reasons. Newer technology may have similar problems. To enable some of the ideas discussed in the application chapters high speed networks are an essential prerequisite, others will only need a modest upgrade in capacity. One of the main areas for improvement will be in the home where communication has largely been neglected, except for the telephone. With increased emphasis on using the home as a work place and as a leisure base the need for better communication is clear, but the service providers also need to ensure that the pricing and facilities are appropriate to the user.

10.5 TRENDS IN SYSTEMS AND APPLICATIONS

The major trends in multimedia communication systems are fairly clear. The communication networks that are available are increasing in capacity as new technological options become available. The systems available over the telephone have improved from 300 bps to 28 800 bps in a few years. This is now pushing the limit of the analogue system that it uses to deliver the data. The use of ISDN has not been as widespread as predicted and it has also taken a long time for connections to become available. Newer technologies do not yet appear to have stabilised sufficiently for any of them to be taken as the clear next step in the upgrade of communication facilities. Although, ATM does appear to have the backing of many service providers and some user groups it also promises a good deal of usable capacity for multimedia communication needs.

There are significant differences in the requirements of users in different situations and the service and network providers do not always provide the most appropriate service for individual users. Home-based users often require high capacity connections to allow access to services such as video-on-demand or to facilitate teleworking from home (although most teleworking can be done with fairly low speed connections). In business the main requirement is to have appropriate capacity for the information that is needed. This is tied closely to the need to design systems so that the information is available to users as they need it without too much duplication and unnecessary communication. These are difficult problems to solve and tend to have different solutions for different businesses. A thorough analysis of the business communication needs is essential if problems, or unnecessary costs are to be avoided.

The same can be said for any application. An information analysis needs to be prepared so that unnecessary technology is not used and what is used allows the users to perform the tasks they are required to do. In education, libraries and the home the various applications available should be designed to perform the functions required. The use of networks and multimedia communication systems is a secondary consideration, but if they are available they should enable more information to be accessible to the user with a usable interface, effective referencing and comprehensive coverage. Few current systems go this far.

The other main area of upgrade is in multimedia workstations or personal computers. The first specification of a multimedia system PC seems so under-powered it is hard to believe that the systems could provide any useful functions but they gave rise to later specifications that were more appropriate and allowed more appropriate use of different media types. The current specifications of PCs with fast processors and internal architectures that are specifically enabled to process multimedia information are still only a step on the way to faster and more accurate processors in the future. This trend is liable to continue as processor chips become faster and hold more processing power. Alongside this the special purpose devices that process information streams such as video and audio will become more integrated within systems and present the user with a complete solution in a box. These systems are technically feasible at present but not widespread. The trends appear to indicate that the systems will become more integrated and incorporate more features.

The final area of improvement is that of applications. Most multimedia tools and applications are still in early generations and have been put together to work with machines whose specification is changing rapidly. A continual catch-up is taking place between the software and hardware providers whereby the software improves to suit the hardware which improves to make better use of the software etc. This all aids the user in the long term but tends to leave them with out-of-date hardware and software as soon as it has been purchased.

There are also limited choices for users at present, with few pieces of software being written specifically for use with multimedia communication networks except for the widespread Internet applications which have evolved but do not yet provide full function software for multimedia networks. There is still some scope for upgrades to systems to incorporate more features and extra media types, both offline and online. Mostly, application software concentrates on individual specific functions and the power of mixed multimedia information services is not realised. One initial step has been taken by WWW browsers which incorporate the use of many different media types although it is at present not a viable proposition to attempt the use of much video on most of the Internet since it does not have a guaranteed capacity for users to rely on. With the advent of higher capacity communication and better accessibility the software will need to improve to incorporate more use of highly data-intensive media, such as video. Many features that are not provided at present may be needed in software to provide users with the facilities to manipulate multimedia information at the desktop, both from their own resources and that available over the networks and from other sources via communication links.

10.6 NEW POSSIBILITIES AND PROBLEMS

The earlier chapters of this book have tried to concentrate on the reality of multimedia communications and its associated technology without trying to predict what might happen if the world is better connected. That is the realm of science fiction. However, there are inevitably some topics ripe for speculation in each of the application areas outlined in the book and it is worth looking at the various possible consequences of wider connection for these areas. There will always be a number of possible options available and the acceptance of one particular implementation of a technology may determine how the world is organised for some time after its adoption. There are also, as has been seen, many other deciding factors such as economic and political decisions which also govern the use of technology in society. These will be discussed further in the next section. The rest of this section will look at each of the three application areas, i.e. business, education and the home, and discuss the possible outcomes and possible problems of implementation of multimedia communication technology.

10.6.1 Business

Although business is seen as being at the forefront of technological innovation it does not always do anything that requires advanced technology. If this changes this will cause major changes to the organisation of work and society. The use of more computers in business allows more automation and higher productivity. This, in turn, removes the need for people to intervene in the production process. So the industrial process has become as mechanised as the agricultural process and people can benefit from greater emphasis on the information processing industries. If all processes were as mechanised as possible most people would need to work in the information industries to provide information and entertainment for other people! There are obvious implications for the whole of society if this were to be the case. For business itself the increased use of computers and communications allows business to be done differently. Easier communication and more automated information processing allied to automated production processes allows more time to be spent on ensuring that customers are served and the goods or services provided are suitable for their purpose or it could allow business to function with fewer staff. Both options are currently employed where new technology is implemented.

Much business now depends on information as a raw material. This type of business is also likely to change radically in years to come. More network connections will allow more access to information for businesses and their employees. They will also allow work to be carried out from different locations. This type of information business is one of the first that are likely to be known as virtual enterprises where there is no real central place that could be called "the office" but the workforce is all teleworkers working from home and the information used is derived from network sources, processed and then used to produce some added value. Consider the example of the electronic newspaper. At present, various news agencies and reporters provide information to newspapers which is then edited into text and images on paper in a central office, but using reports from around the world. In the electronic scenario the office could disappear, the newspaper would become electronic, available over the network. The reporters, editors and other staff could work from home compiling reports, advertising etc. and the product could be sent to subscribers over the networks. Much the same organisation is needed to produce the newspaper as before but the work could be done via multimedia communication networks rather than having people travel to offices produce information, print it onto paper and then distribute it via road or rail to newsagents. An intermediate step might be to produce facsimile printings in various centres around the country to reduce transport costs. These would probably need to be shared between information providers to justify the cost. There are many possible scenarios for the gathering, editing, production and distribution of a newspaper. These range from the present reliance on paper to a fully networked solution. There are a number of intermediate possibilities for all the stages of obtaining an end product.

Much of the advantage of networked delivery of information products relies on the public acceptance of information provided in this form. While the users of computers are often seen to be requiring more extensive access to information they are still only a minority of the information users in society. Many users will still require information on paper for a number of years and any information provider wishing to extend their market opportunities would need to concentrate on both electronic delivery and more efficient paper-based distribution at the same time. The electronic newspaper may be an interesting idea for computer users but it is inaccessible to general users of traditional paper products. A possible solution is to provide an alternative delivery mechanism of either paper as it is used at present or some other form of paper delivery such as just-in-time printing in newsagents or similar outlets. There are many possible outcomes for the deliberation of system implementers in future.

Other information providers may use alternative delivery mechanisms should they become widely available. The rapidly spreading wiring of homes by cable TV network operators allows more data into the home than previously and this gives information businesses more opportunity to sell products aimed at the domestic market through added services such as database access, home shopping etc. Much of this will depend on set-top box provision and the ability of the networks to provide the bi-directional capacity required by the services. Future directions of information provision to domestic consumers can only be speculation at present. The opportunities to target homes is one that is unlikely to be missed by information providers and the services offered are likely to be as diverse as the range of services currently offered over the telephone, but with the added value of multimedia information.

10.6.2 Education

The world of education has taken to multimedia communication quite rapidly and would appear to be one of the major areas for the growth of individual multimedia products and services for many years to come. Since education depends on information and communication it is only natural that the use of multimedia communication would be widely applied to educational needs. Indeed, many different media have been used in the past to enhance the educational experience. Some have been more successful than others. The possibilities are, however, very diverse and almost endless in scope. Information for education can be provided via networks easier than with paper-based formats allowing easier updating and more flexibility for teachers and learners alike. Schools, colleges and universities can all plug into the latest information resources and give students the best, most up-to-date information for learning. There are problems with this. The users need to be able to access information in a coherent form and this will require some guidance, probably from a teacher. There are, however, few teachers at present that could provide such guidance. Information needs to be provided in a suitable form for the users and to be available so that the users and their teachers can access it easily.

One possible consequence of such an approach would be in the growth of independent learning. More access to information and material on networks could allow learners to guide their own learning experience with consultation with any educator on the network world-wide. This would reduce the dependence on formal structures of education such as schools and allow greater diversity of learning for individuals. It would not give any of the informal learning that schools and colleges provide in the socialisation of students and this would probably be to the detriment of society since this function would be passed down to individual homes and all the variety encountered there. The networks can enable any number of scenarios to occur, some may be little different from current practice, others may be widely removed from experience. The function of learning is often seen as an area for popular debate and it will inevitably remain so, but if multimedia communication networks are used for the delivery of the educational experience the consequences should be explored first. Greater access to education would ensue from wider connectivity. This would also lead to wider use of the networks by non-educational sources such as commerce and industry. Much positive information could be used from such sources but there may also be negative consequences of this type of interaction with the educational process. Education has, for a long time, tried to involve outside concerns with the learning process and this has been successful in places where companies and other bodies enhance the process. The interaction has largely been controlled by teachers. Wider accessibility to education by business and other interests may not change the nature of the interaction, but there are possible areas that may distort the learning process of individuals, especially in an independent learning environment.

Further network use, by educational users, is inevitable with the scarcity of traditional resources and the ease of access to world-wide information that is available from networks at present. Future developments can only enhance the process and provide faster, more accessible information to learners and teachers alike.

10.6.3 Home

The home is probably the main area where multimedia communication networks will have most effect. There is more information used in the home than is generally realised and many of the current developments are directed specifically at the home information user. The technology available for use in the home will enable masses of information into the home from diverse sources with less control over content than is currently afforded by the intermediaries in the information process, e.g. editors and producers. More access to information via multimedia communication networks also has the effect of diluting the unity of information as would be presented by traditional sources. More diversity can be encouraged, or rejected by the individual user and information providers need not provide information in the same format as traditional media since the links between different media types can be dynamic and different media mixes can be used for any information provision. The home has been a user of diverse information sources for many years. Users often access information from newspapers, television and other sources and use the different sources in different ways, each having its own relative strengths and weaknesses.

There are also implications for the organisation of the home. The possibility of better communication between distributed people can make homes more accessible to members of the same family, removing the divisions imposed by place upon the communication that takes place between them. This aspect, coupled with better access to information in all its forms allows new paradigms of living to be postulated within homes and society as a whole. Many different effects could come out of this and other changes. What may be a brake to expansion in this area is the loss of privacy or the increase in physical isolation that these ideas may involve. Society has grown into what it is by virtue of the people in it. Changes brought about by technology do not necessarily change what people do just because it is possible. It may be that there is no interest in such possibilities and society continues as it has before with extra communication channels. There are, however, precedents to show that there may be an eventual change in the way society is organised. The growth in technology in the twentieth century has changed society by changing the nature of work and family structures. There is every possibility that these changes will continue with the use of multimedia communication technology. There are no effective ways of predicting what these changes may be, but the increased use of telecommunications may lead to more isolation at home and at work, especially if teleworking is more widely used and homes are used more for leisure activities.

Further speculation leads to some possible organisations of society that are dramatically different from today's picture. However, that are many problems with speculation and it is too early to predict much of what society will become. There are, though, some ideas to bear in mind as technology is introduced and small changes occur since even the introduction of simple changes in technology can have far-reaching effects.

10.7 SOCIETAL CONSIDERATIONS OF MULTIMEDIA COMMUNICATION TECHNOLOGY

A number of previous chapters and sections have indicated that there are many factors, such as economic and political decisions, other than the technology itself which also govern the use of technology in society. Some of these influences on the information society are discussed in the following sections. Again this section is largely speculative in that the technology is not yet sufficiently widespread to show trends in any particular direction, but some of the influences on previous technological implementations will still be relevant to new technological developments. These influences are political and economic and concern the aspects of access to, and control of the technology.

10.7.1 Politics

Political influence can never be ignored in any application of technology on a wide scale. The use of standards in communication has arisen out of co-operation between telecommunications companies who have largely been subject to political influence as public corporations (especially in Europe). The large scale privatisation of the same companies has changed some of this co-operative working into competition. Other industries, such as the computing industry have never been subject to the same restrictions and have, largely, eschewed standardisation in favour of individual solutions. Political influence extends into many corners of life. The construction of information superhighways is partly dependent upon the support of government to underwrite the operation. The installation of cable TV networks has only been possible in many countries by a change in the law allowing operators to provide services. This, in turn, should lead to information superhighways connecting together the various end-point providers. This is not necessarily a priority of the providers but can be influenced by political leaders and this has been shown to be the case.

Other areas of political influence may not be so straightforward. The whole political environment determines the ease with which any information society technology will be developed. If conditions are not suitable for a large-scale adoption of the technology it can be highly unsuccessful. The two cases of the UK and France and the introduction of the similar services of Prestel (UK) and Minitel (France) are good examples of the role of political influence. Prestel provided for access to information but at a cost to the consumer that has largely disappeared, whereas Minitel has developed on the back of large government investment in a supportive infrastructure. Lack of government interest led to Prestel being a specialist service from the beginning, whereas the Minitel terminal has found much wider acceptance.

The same is likely to be true of multimedia communication networks. A number of governments have already shown support for large-scale infrastructure projects to support an information society and these will need to be extended to enable many of the possibilities outlined in this book. For an initial impact many homes and businesses will need to be connected together to enable greater exchange of information. This may only come about with government encouragement, especially if the economic signs are against such an investment. The subsequent society formed by greater connectivity will also be subject to political influence and its nature will, in part, be determined by the emphasis of government. For example, a government that stresses the importance of the information society solely to business will not be in a position to aid its extension to the home which is likely to be a more profound area of application. Political influence is widespread and many of the other areas that impinge upon the implementation of technology are not free of political influence.

10.7.2 Economics

Many of the other influences on technology are largely irrelevant until the economics of any implementation allow it to be adopted by users. Many technologies could help business, education and home users to increase their access to information and provide better tools for work and study but are, as yet, uneconomic to deliver within the constraints of a limited budget. This can often be the case for new, and untried, technology. At present, high speed network connections are priced too high for ordinary users and do not allow many access to this technology. In time, prices will fall and more users will be able to install the technology. The case of ISDN has already been discussed, but with lower pricing, much more widespread take-up would occur having a knock-on effect on the prices of related equipment. This situation has occurred again and again with different technologies, e.g. telephone, fax, PC, calculators, mobile telephones etc. The economics of any technology determine its initial impact, careful pricing determines its later take-up. Multimedia communication is likely to be the same. Initial equipment has been expensive and later developments have produced more functions for less outlay. Eventually it will become cost effective to include the hardware in new PC packages. this will lead to greater uptake and wider spread of the technology to other interested users. This largely assumes that there is sufficient interest in the services offered by these networks to enable the critical mass of users to form and for the technology to make an impact. There needs to be many different factors acting at the same time for a successful implementation.

10.7.3 Access to information

The success of any information-based society will depend on the access to information that is possible with the systems and networks that are installed. To function effectively a business or an educational establishment needs to be able to obtain relevant pieces of information to its function. To do this may require large-scale connection to the networks or continuance of current methods and modes of operation for longer than is strictly necessary. For example, if a college wished to offer courses via distance learning with full online resourcing it would need to ensure that the library facilities needed were available online or ensure that the resources could be provided in some other way. To function effectively in an information society needs widespread connectivity. Then homes, businesses and other institutions will be able to share information and provide services between each other.

In addition, the information that is made available needs to be accessible. This is often problematic with information being stored in different forms by different providers without using uniform access mechanisms as have been available for information in traditional forms. It is little use providing information for users if it takes a long time to use it and is difficult to find. Access mechanisms need to be designed to suit the environment of their application. For example, large-scale ftp archives can be very useful for distributing program updates, shareware or documentation, but without relevant indexing and organisation they can be practically impossible to use.

10.7.4 Access to systems

There is also the problem of access to the systems used to access the information. Many interesting and useful schemes for the use of computers in homes or business can be proposed but will never see much success because the intended users do not have access to the necessary machines. This limits the usefulness of the networks and the applications and reduces the take-up of applications. This in turn increases costs and makes it even less likely that the intended users will be able to access the systems and the information. For example, the use of electronic data interchange has been spreading in industry for many years. Its spread is very slow with many companies, especially smaller ones, reluctant to invest in a technology that is relatively expensive (compared with their turnover and profits). This leads to a situation where most users of EDI are large organisations with much larger budgets. This tends to increase the cost of software solutions and deters smaller businesses from further investigation of EDI.

The situation in the home is similar. Access to information via the television and newspapers is a cheap and simple solution. Introducing computers is fraught with difficulties. Firstly, the economics are against their widespread use since the average household does not consider the use of a computer as a priority issue. Secondly, their use is still not as simple as computer companies would like us to believe. To many potential users a computer is of little intrinsic use and other solutions may provide better access to information than the current design of home PC.

10.7.5 Control in the information society

One of the major issues that has come to light in recent years is the problem of control of information. This can be both in the sense that information can be controlled by individuals and also that there can be a lack of control of unsuitable information for vulnerable clients, such as children. The first problem has been a feature of information since the first mass media were invented and used. The control of information being in the hands of the proprietors of the mechanism for publication and distribution which has largely been newspapers and more recently television channels. To a certain extent the use of multimedia communication networks can ameliorate this effect if the user wishes to exert their own control over the information obtained. This does, however, require effort and is unlikely to be as common as the easier option of accessing single sources of edited information material. The second problem of lack of control is a current talking point for the users of the Internet. There is, currently, no real solution to those users who take advantage of the networks to distribute offensive material as there are many ways this can be done. Some of the more public arenas can be removed but they are likely to lead to less public methods of distribution and these will still attract the same audience as at present. There are no simple solutions especially since the networks are global in nature and often the activities may be illegal in one country but not in another.

Control of multimedia communication networks is a problematic area and there are few real tried and tested solutions. Many suggestions are made and few have any basis in reality, often being reactions to specific events or incidents. The future of networks can largely be said to depend on any solutions to this and the other problems outlined here. Without a comprehensive policy on access and control of information the networks are unlikely to develop beyond the current paradigms and furnish the world with an information society.

10.8 FINAL SUMMARY

There are many commentators that have written about the information society and two of these were referenced at the end of Chapter 9. They give different views of what it might mean to be in an information society and are both valid views. The information society is an idea that has gained momentum over the last few years and multimedia communication is one of the technologies that will make or break the visions of such a society. The technology alone can be used to provide more effective communication between people, between businesses, and between people and organisations working together. Without multimedia communication the information society will not be much different from that of today, with multimedia communication it is hard to predict the scale of the changes that could happen. There are still problems that need to be solved, some technical, some organisational and some of a more human nature. There is still much to learn about the organisation of society when it can use multimedia communication and some aspects of the process may prove difficult. However, there is potential for much benefit to accrue to society and the opportunities need careful consideration. These benefits are still to be fully determined, but the process has started and some changes will inevitably occur. To influence the changes requires some understanding of the technology and its potential.

EXERCISES

1. Discuss in what ways multimedia communication networks can be used to serve healthcare applications.

2. Discuss the different ways in which linkage occurs between medicine and the other application areas covered in earlier chapters.

3. What are the influences on the installation of optical fibre to the home?

4. Discuss the control of information available over the networks available in your own country.

GLOSSARY

Acknowledgement: A response to a communication. It can be positive or negative, i.e received OK or in error.

Address: The unique identifier for a person or machine within a network.

Aloha: A radio broadcast protocol invented in Hawaii around 1970.

Amplitude: The size of a wave or signal. This equates to the volume of sound, the brightness of light, or the strength of a radio signal.

Amplitude Shift Keying: A modulation technique based on using different amplitudes to represent different symbols. For example two amplitudes to represent 0 and 1.

Analogue: A representation of a physically varying quantity using a continuously varying wave. For example an electrical signal to represent sound.

ASCII: American Standard Code for Information Interchange. A coding system for text characters using seven bits per character. A national variation of International Alphabet No. 5.

Asynchronous: Signals that are sent with embedded information for synchronisation purposes. For example characters using start and stop bits. The extra bits allow the receiver to synchronise itself for the data being received.

Asynchronous Transfer Mode: A modern high-speed data network protocol that uses 53-octet cells to transfer information in a connection-oriented configuration. Seen as the ultimate replacement for many network types at present.

Attenuation: The loss of strength of a signal caused by various factors. Often due to unavoidable transmission characteristics.

AT-command language: The most popular modem control language. Allows commands prefixed by the characters AT to control all modem functions.

Automatic repeat request: A form of protocol for error control on duplex channels. The repeat of erroneous data is requested by negative or no acknowledgements.

Bandwidth: The range of frequencies that can be used in a channel which, therefore, ultimately determines the capacity.

Baseband: A signal sent without modulation as direct digital levels.

Baud rate: The number of changes in a signal per second. Often the same as data rate, but not with multi-level signalling.

BCH codes: A group of Forward Error Correction codes which includes the Hamming code as a special case.

Binary digit (Bit): A fundamental digital quantity, represented by 0 or 1.

Bit stuffing: A technique used by X.25 and HDLC to ensure data transparency. Extra bits are inserted into certain sequences to force them to be different from control sequences.

Block check code: The code used at the end of a frame of data that is composed, in some way, from the data itself. When the same mechanism is used by the receiver an error can be detected if the calculated code differs from the received code.

Broadband: A system that uses modulated channels to increase capacity.

Broadcast: A technique used in radio and LANs where a data unit is sent to all possible recipients indiscriminately. The intended recipient must check the data for its own address.

Burst error: A burst of noise that affects a number of sequential bits of a communication.

Byte stuffing: A technique used in character-oriented protocols to ensure data transparency. A special character is used to indicate that a character is not to be used for control purposes.

Byte (octet): A group of 8 bits.

CAD/CAM: A combination of computer aided design and computer aided manufacture. Engineering drawings have a large memory requirement.

Capacity: The amount of data that can be transmitted through a channel.

Carrier: A wave used for modulation of a signal.

Cellular radio: A network of radio stations used for two-way mobile telephone communications.

Channel: A single message path through a medium. A medium can normally allow many channels simultaneously.

Coaxial cable: A cable which consists of a central conductor surrounded by an insulator and an earthed shield, used in bus-type LANs.

Coding: The representation of information for transmission.

Communication: The effective transmission of information between sender and receiver.

Complex parity: A combination of parity bits to increase error detection and correction capabilities.

Computer system: A electronic system for the manipulation and storage of data. This can be a personal computer or a mainframe.

Congestion: A condition caused when part of a network has more traffic than its capacity.

Context: The background that allows extra information to be carried implicitly in a communication.

CSMA/CD: Carrier sense multiple access protocol with collision detection. Used in bus-type LANs.

Cyclic redundancy check codes: A code word calculated from the data in a communication and sent along with it. It allows many different sizes of burst error to be detected.

Data communications equipment: The generic name for the communications side of the interface between a computer and the PSTN.

Data compression: Schemes for reducing the redundancy in data by using different techniques for representing repeated bit patterns.

Data rate: The amount of bits sent through a channel per second.

Data terminal equipment: The generic name for the computer side of the interface between a computer and the PSTN.

Data transparency: The ability of a protocol to transmit any bit pattern.

Data unit: The format used for sending information within a protocol layer.

Database: A structured file or set of files used to contain related information.

Datagrams: The method of sending data through a network that has address information attached to each packet of data, allowing different routes to be taken by different packets.

De facto standard: A standard established by popular use.

Digital: The representation of information by discrete levels rather than infinitely variable amounts.

Distortion: The change in a signal due to variable propagation conditions.

Domain: A group of related addresses in a network. For example a PRMD in X.400.

Duplex: A two-way communications channel.

EBCDIC: Extended Binary Coded Decimal Interchange Code, an alternative character code to ASCII.

EDIFACT: The international standard for EDI messages.

Efficiency: The ratio of total data transmitted to actual data transmitted.

Electronic Data Interchange (EDI): The exchange of business documentation by electronic means.

Electronic Mail (Email): The electronic exchange of messages between computer users.

Error correction: The ability to correct an error that has occurred in transmission.

Error detection: The ability to detect but not correct an error that has occurred in transmission.

Fax: The electronic transmission of documents and images using facsimile techniques.

Fibre: See optical fibre.

File server: A computer attached to a network that is used for the storage of files for the network users.

Flow control: The ability of a protocol to allow for different speeds of sending and receiving equipment.

Forward error correction: The technique used to correct errors by adding extra data to a transmission to increase redundancy.

Fourier series: The representation of a waveform as a combination of sine and cosine waves.

Frame: A combination of bits that are used to outline data in a transmission. Can include address and control information.

Frame check sequence: The data word calculated from the data in a frame to allow error detection. The FCS is re-calculated on reception and this is compared with the received FCS data word.

Frame relay: A new form of WAN protocol that allows high speed transmission.

Framing: The addition of extra bits or bytes to form standard data patterns for reliable transmission.

Frequency: The number of cycles of a wave per second.

Frequency Shift Keying: A modulation technique which uses different frequencies to represent different symbols.

FTP: File transfer protocol. The Internet protocol for file transfer.

Gateway: The means of converting protocols between different networks. Usually implemented on either a stand-alone machine or a file-server.

Graphics: The computer representation of diagrams and images.

GSM: The European digital cellular telephone standard (also in use in other countries).

Half duplex: A duplex communication channel which can only be used in one direction at a time.

Host: A computer attached to a network via a node. Often host and node are the same machine, but originally they were separate.

IEEE 802: The IEEE standards for local area networks.

Image: The digital representation of a picture.

Information: An abstract quantity that has some meaning. A combination of data and encoding rules with context and other aspects that render data meaningful to a receiver.

International Direct Dialling: The ability to dial a telephone number for any subscriber in the world without operator intervention.

International standard: A standard decided by some international body.

Internet: The combination of various networks that use the Internet protocols and addressing rules.

ISDN: Integrated services digital network. A fully digital network that allows direct digital connection to the subscriber.

ITU-T: An international standards body for telecommunications standards.

Leased line: A private circuit leased from a public service provider for exclusive use.

Local area network: A network of computers that covers a small geographical area, such as a room, building or a set of buildings occupied by one organisation.

Logical view: The user's view of a system, which can be presented as different from the physical view by software.

Mainframe: A large computer system. Multi-user, multi-programming machine that can service many simultaneous users and programs.

Medium: The physical means used to transfer signals. For example radio, cable, etc.

Message: The information contained in a communication. Encoded for transmission in the medium.

Message format: The structure of the message. Used as part of the overall protocol.

Minicomputer: A medium-sized machine, between mainframe and desktop computer.

MNP: A set of data compression protocols used with modem communications.

Modem: A device that allows connection of a computer to the PSTN short for modulator-demodulator.

Modulation: A technique for transmitting digital data using analogue signals.

Modulo arithmetic: Arithmetic that is carried out on finite number sets. It is cyclic in nature.

MS-DOS: Microsoft's disc operating system. The most popular PC operating system.

Multi-level signalling: The use of more than two levels per signal change to increase data rate.

Multi-user system: A computer system that allows simultaneous use by many different users.

Network: A set of connections between a number of computers.

Node: A machine on a network that is used to send and receive information.

Noise: Parts of a received transmission that were not included in the original communication introduced by external factors.

Open Systems Interconnection Reference Model: An international standard framework model for computer communication.

Optical fibre: A medium of glass 'wires' that allow the transmission of information using light signals.

Parallel data: The representation of data that uses a number of different simultaneous signals. A number of different wires, each one used for one bit of a data word in data transfer.

Parity: The exclusive-OR sum of the bits in a data word (even parity) or its inverse (odd parity).

PBX: A private telephone exchange used by a business.

PC-personal computer: A small computer system that is frequently used by only one person.

Phase: A fundamental property of waves. Best described by the difference between waves that are of different phase.

Phase Shift Keying: The use of phase change as a modulation technique.

Phoneme: A fundamental symbol used in speech.

Physical view: The actual layout of hardware and media. Can be presented to the user as a different logical view.

Pixel: A pel with colour or different shades of grey.

Protocol: The set of rules that govern a communication.

Protocol layering: The use of different protocols for different purposes. Each layer is embedded in the layer below ending in an encoded bit stream.

Public Switched Data Network (PSDN): The data equivalent of the telephone network. Usually an X.25 network.

Public Switched Telephone Network (PSTN): The worldwide telephone network.

Quadrature amplitude shift keying: A combination technique for modulation using amplitude and phase changes to represent data.

Quadrature phase shift keying: A modulation technique that uses phase changes to encode data.

Radio: Electromagnetic waves used as a medium for wireless communication.

Recovery: The recovery of a communication after channel or equipment failure.

Redundancy: The extra data that is used to ensure the accurate transmission and reception of information.

Reed-Solomon codes: Another set of forward error correction codes.

Register: The fundamental storage unit of a central processing unit.

Remote LAN bridge: A means of connecting two parts of a LAN that are separated by a larger distance than can be covered using standard LAN cabling.

Re-tries: The re-sending of a communication when no acknowledgement is received.

Routing techniques: The schemes used to send a message through a network.

RS232: A standard interface specification. Originally for computer to modem connections.

Sequence control: The ordering of data in a communication to recover the original message when a message is sent out of sequence.

Serial data: The transfer of information one bit at a time.

Server: A computer attached to a network that is used to perform a particular task or set of tasks, For example a file server or print server.

Shannon-Hartley law: A measure of channel capacity for AWGN channels.

Shared data medium: A medium shared by a number of communicating devices. For example a bus-type LAN.

Signal: The means of conveying a message in a medium.

Simple Mail Transfer Protocol: The Internet mail protocol.

Simplex: One-way communication.

Sine wave: A regular analogue wave generated by circular motion.

Skew: The phenomenon caused by the arrival of parallel signals which are out of phase.

Spreadsheet: A software package used to perform simple calculations on data in the form of a screen with cells with associated formulas.

Standards: The agreed format and specification for objects and services.

Store and forward: A network system that sends messages between nodes by storing them before re-transmitting them towards their destination.

Sub-network: A network that acts as part of a larger network but is also self contained.

Switching techniques: The techniques used to determine the type of connections used for a communication.

Symbols: The encoding system used for representation of information.

Synchronous: A data transmission that is permanently synchronised during transmission.

Telnet: The Internet remote login protocol.

Time-out: This occurs when a transmitter does not receive an expected acknowledgement within the specified time.

Token passing: The medium sharing mechanism that uses a token that is passed between nodes.

Topology: The physical configuration of a network.

Twisted pair: A medium consisting of a pair of conductors twisted around each other.

UART/USART/USRT: Programmable electronic communications devices.

Unix: A popular minicomputer operating system.

Value added data services: The term used for network providers that also provide other services such as EDI mailboxes.

Virtual circuits (Virtual channels): The means of determining the path of data in a packet switching network by choosing the path in advance.

V-series recommendations: The series of ITU-T recommendations concerned with physical interfaces and data transmission speeds.

Wavelength: The inverse of frequency. The length of a wave in a medium.

Wide area network: A network that connects nodes that are separated by long distances.

Windows: A graphical user interface for personal computers.

Word processor: A software package that assists with the preparation of text and documents.

X-series recommendations: The series of ITU-T recommendations that are concerned with the predominantly logical aspects of protocols.

X.25: An international standard network access protocol.

X.400: An international standard for electronic mail.

X.500: The international standard for directory services.

INDEX